T0160279
Archangel
Christiania
St Petersburg
Moscow
Berlin
Warsaw
Vienna
Constantinople
Sea
Teheran
Cairo
Mecca
INDIA
Bombay
Madras
Malabar Current
Equator
INDIAN
OCEAN
MADAGASCAR
Cape of Good Hope
Sea Weed

LIFE LESSONS
FROM EXPLORERS

Published in 2021 by Welbeck

An imprint of Welbeck Non-Fiction Limited, part of
Welbeck Publishing Group
20 Mortimer Street
London W1T 3JW

A catalogue record for this book is available from the British Library.

ISBN: 978 1 78739 611 1

10 9 8 7 6 5 4 3 2 1

Printed in Dubai

Editorial: Isabel Wilkinson
Design: Russell Knowles, James Pople
Picture Manager: Steve Behan
Production: Marion Storz

LIFE LESSONS

FROM EXPLORERS

LEARN HOW TO WEATHER LIFE'S STORMS FROM HISTORY'S GREATEST EXPLORERS

FELICITY ASTON MBE

WELBECK

CONTENTS

INTRODUCTION

I had made a critical mistake. Skiing alone across Antarctica during an expedition to traverse the continent's landmass, I made my way up a steep and narrow glacier that formed a rare navigable path through the Transantarctic Mountains. No one had skied up this glacier before, so I had little idea of the terrain ahead of me. Winds from the elevated central plateau of Antarctica were being forced down the glacier towards the coast and, having spent a gruelling day inching upwards into those punishing headwinds, I eventually decided I had no choice but to pitch my tent and get some rest. I spent a few tormented hours trying in vain to sleep, being lashed by tent poles bent so low by the wind that I was terrified they would snap, and being forced to venture outside into the brutal Antarctic cold every 60 minutes to heap more snow on the fixtures that kept my precious shelter anchored to the ground. With my tent hopelessly exposed on the bare face of the wind-scoured glacier, I gave up on the idea of rest and laboriously dismantled my camp to wearily move on.

Not more than a few hours' toil up the glacier, I discovered there was a wide bowl in the ice, protected on all sides from the destructive wind. If I had continued just a little longer the previous day, I would have been able to camp in this perfect spot and get the respite I so desperately needed. I was furious with myself for the misjudgement – but the experience gave me a glimpse of the frustration, uncertainty and burden of what it means to be "first". I thought of those early explorers who, a century earlier, had pioneered routes through this most extreme and remote of mountain chains in search of the South Pole beyond, and gained a new perspective on the anguish of the decisions they had to make in the course of those journeys that would become legendary. I was also struck by the difficulty, when equipped with the hindsight of history, to accurately imagine and appreciate the true complexity of their experiences. Is it possible to conceive the weight of responsibility Sir Ernest Shackleton must have felt as he contemplated leading his men when they were far from any possibility of rescue, shipwrecked on fragmenting ice? Can we ever truly conjure the emotional challenge of diplomatically ingratiating ourselves into a tribe of armed and quite possibly hostile Bedouin in the solitary desert, as Gertrude Bell dared? Or having the self-discipline to count every step of a relentless trek across the roof of the world for years as did Nain Singh Rawat?

Yet it is important to remember that none of these legendary explorers was vested with superhuman powers. They were each human, with their unique flaws and talents, just as we all are. While this fact only makes these heroes even more remarkable and fascinating, it is also a reminder that every one of us has the potential to be just as extraordinary and achieve feats as great as those of the explorers featured in this book. Here, the purpose of my examining and retelling the successes – and often more importantly, the failures – of some of history's greatest explorers is to unlock the innate abilities within us all. My intention is that you are each inspired to let your ambitions soar as high as Amelia Earhart, to have the joyful determination of Junko Tabei, to overcome adversity with the hope of Olaudah Equiano, and to value humanity with the same conviction as Neil Armstrong – not in order to emulate these trailblazers, but rather as a way of uncovering the very *best* version of *you*.

ADAPTABILITY

ROALD AMUNDSEN

IN THAT THEY ARE BOTH A COLD, WHITE WILDERNESS OF ICE AND SNOW, THERE APPEARS TO BE LITTLE DIFFERENCE BETWEEN THE OPPOSING POLAR REGIONS: THE ARCTIC AND THE ANTARCTIC. YET THE TWO ENVIRONMENTS ARE, IN FACT, COMPLETELY DISTINCT FROM EACH OTHER. THE ARCTIC IS A FROZEN OCEAN, A HUMID REGION OF PLENTIFUL LIFE AND ANCIENT HUMAN CULTURES. THE ANTARCTIC IS A CREVASSE-SCARRED DESERT OF ICE, A STERILE CONTINENT WHERE SMALL CLUSTERS OF HARDY MICROBES ARE THE ONLY LIFE FORMS.

In September 1910, a squat wooden ship called the *Fram* was harboured in the port of Funchal on the Atlantic island of Madeira. On board were 19 men, 100 Greenland sled dogs and several years' worth of sledging rations and supplies. The ship had reached its first port of call on its way to the Panama Canal and across the Pacific Ocean to the Arctic, where it intended to reach the North Pole by being frozen into sea ice and transported over the pole by the drift of wind and current. Just before the ship was due to leave Funchal, the expedition leader, Captain Roald Amundsen, gathered his crew to make an astonishing announcement. There had been a change of plan. The *Fram* would be heading not northward to the Arctic, but south to Antarctica instead. The new mission of the expedition was to be the first to reach the South Pole.

The men were initially apprehensive about this abrupt volte-face. The equipment prepared for Arctic travel had not been proven in Antarctica, and those few expeditions which had attempted to venture south had decreed the terrain unsuitable for sledging with dogs. Most troubling of all, a large, government-funded and well-equipped British expedition

Amundsen (far left) and his crew in 1906 aboard the small but robust sealing vessel, Gjøa, having arrived in Alaska as the first expedition to successfully navigate the Northwest Passage.

had departed for Antarctica three months earlier, to much fanfare, with the clearly stated aim of reaching the South Pole.

Amundsen reassured the crew of their preparedness for Antarctica by detailing his meticulous planning. He revealed that the decision to travel south had been made more than a year beforehand but had necessarily been kept a secret from everyone, including the crew. Indeed, Amundsen felt that the change of direction was one that had been forced on him. Almost two years earlier, in November 1908, Amundsen had made public his plans for the *Fram*'s North Pole expedition – only to be dealt a huge blow five months later when the American explorer Robert Peary announced that he had successfully reached the North Pole, claiming the record of being first on which Amundsen had set his sights. Around the same time, news broke that the *Nimrod* Expedition led by Sir Ernest Shackleton had failed to reach the South Pole, despite coming very close. Amundsen then determined that, after missing the chance to be first to the North Pole, he would claim the remaining accolade in the south instead. He had not made his decision public because the funding for his *Fram* expedition had been awarded on the basis of conducting scientific research in the Arctic, not the Antarctic, and because he was worried he might be refused permission to enter into direct competition with the British expedition to the South Pole, which had already been announced.

The British Antarctic Expedition, led by experienced Antarctic explorer Captain Robert Scott, had all the advantages of a major national undertaking. It had been advised by the most accomplished polar explorers of the day and was equipped with cutting-edge technology, including motorized sledges that were expected to revolutionize travel in Antarctica. Amundsen told his expedition team aboard the *Fram* that he meant to contend for the pole against Scott's larger expedition and that he expected that they, being small in number, and lightweight and quick in comparison, would reach it first. His argument won over the crew: the combination of agility and efficiency was a strategy that had worked well for Amundsen before.

As a boy, Amundsen had been fascinated by the story of Sir John Franklin. In 1845, the 59-year-old British veteran of the Arctic had been in command of two elaborately refitted warships featuring all the latest technological advancements, a compliment of 133 men and some 13,600 kg (30,000 lb) of food. His expedition was charged with the discovery of the Northwest Passage, a sea route from Atlantic to Pacific via the Arctic Ocean which would replace the need to round treacherous Cape Horn at the southern tip of South America. His was the nineteenth expedition in 300 years to go in search of this fabled northern shortcut and was deemed too big to fail. But Sir John Franklin and his expedition promptly vanished into the Arctic, never to return

– their fate remaining an enduring mystery for generations. Half a century later, a 30-year-old Amundsen set off to achieve his boyhood dream to accomplish what his hero had failed to do: discover the Northwest Passage. In contrast to Franklin's expedition, Amundsen sailed not in a deep-water warship but in a small and battered 30-year-old sealer that had already proven its worth over decades of work in Arctic waters. He reasoned that the smaller ship would be able not only to go through narrower channels between perilous ice floes, but also to navigate shallower seas, leaving the expedition with a greater range of possible route options around the myriad islands and treacherous straits of the High Arctic. Furthermore, rather than hundreds of men, his smaller ship needed a crew of just seven. "What has not been accomplished with large vessels and brute force I will attempt with a small vessel and patience," he promised. This radical approach had been inspired by another Norwegian, Fridtjof Nansen, who had crossed Greenland five years earlier, seemingly with spectacular ease, by pioneering a nimble and lightweight style. Amundsen perceived that where attempts to overpower the Arctic environment with an invasion of modern technology had failed, he would instead succeed by learning how to adapt.

Resolved that he would learn from native Arctic peoples how to overcome the challenges of the climate and landscape, Amundsen followed the route of the Franklin expedition to King William Island, where he planned to spend the winter conducting scientific work on magnetism and establishing the position of the North Magnetic Pole. The party soon made contact with the Netsilik, a local people who moved with the seasons and subsisted during the winter mainly by hunting seals on the sea ice. Over the next 22 months, Amundsen dedicated himself to learning all he could from the Netsilik. He adopted the fur clothes they wore, practised building shelters from snow, and became proficient at travelling with dog-hauled sleds. Amundsen

Amundsen's South Pole expedition forging a perilous route up what the team later named Devil's Glacier due to the tortuous terrain and terrifying crevasses, which had to be crossed using precarious snow bridges.

believed that gaining knowledge was the key to greater adaptability – a principle he had held all his life.

As a young man reading accounts of famous explorers and their exploits, he had noticed that many leaders of expeditions by sea lacked the ability to command their own vessels and relied heavily on the ship's captain. Recognizing that this created a problematic conflict of authority, Amundsen spent three years as a sailor and two years as a ship's mate in order to gain all the knowledge he needed to become an expert navigator and to command his own ship. On King William Island, however, his dedication to expanding his knowledge from the Netsilik almost cost him his expedition. He had failed to share his vision with his crew, who didn't understand his apparent obsession with the native peoples. Worried that their leader was becoming increasingly estranged, they forced him to refocus on their primary objective. On 13 August 1905, Amundsen finally agreed to continue the expedition's journey and three weeks later on 26 August 1905, when they spotted the sail of a ship which had "San Francisco" painted on its side, they knew they had become the first to transit the Northwest Passage.

As he sailed towards Antarctica on the *Fram* in 1910, Amundsen was convinced that the adaptability he had learned from the Netsilik would give him an advantage over Scott – but his previous experiences in Antarctica also weighed on his mind. In 1897, Amundsen had signed on as first mate with the Belgian Antarctic Expedition under the command of Adrien de Gerlache. The expedition became the first to spend a winter in Antarctica (some claim accidentally), but the undertaking had been poorly equipped, lacking even enough warm clothing for the crew, and carried ill-suited rations that quickly led to lethal outbreaks of scurvy. Two men died as the suffering led to terrifying mental illness. Learning from the mistakes he had witnessed first-hand, Amundsen was tireless in his constant quest for improvement. As the *Fram* set up winter quarters at the Bay of Whales on the Ross Sea coast of Antarctica, he oversaw the laying of generous depots of rations and supplies along the route he would take to the South Pole, allowing plenty of provision for bad weather and delays, always preparing for the worst possible scenario. Throughout these depot-laying journeys Amundsen paid close attention to any problems encountered with either equipment or men, adapting his plans to what he was learning about the environment and the capability of those on his team. He used the necessary confinement of the winter months as an opportunity for the crew to work on the design of ski boots that hadn't performed well, modify sledges to eradicate any excess weight and make adjustments that allowed the sledges to run smoother over the ice. No detail was too small to be worthy of time and attention. His focus on achieving the expedition's goal was ruthless, and for Amundsen this meant maintaining his authority as expedition leader as carefully as he did the supplies. When one of the most experienced expedition members openly questioned Amundsen's decisions and accused him of "panic", the man was swiftly separated from the group and dispatched to explore some distant terrain with two companions.

Many uncertainties haunted Amundsen despite all his meticulous preparations. He did not know if it would be possible to get the dog teams he relied upon through the steep and tortuous terrain of the Transantarctic Mountains to the elevated altitudes of Antarctica's central plateau, where the South Pole was located. Most of all, Amundsen feared the capability of Scott's motorized sledges. We now know, of course, what Amundsen didn't: that the motorized sledges were not a success. Furthermore, Scott was placing a far greater emphasis on scientific explorations of the region than on his own attainment of the South Pole – and so taking a complicated and altogether slow and steady approach that was in direct contrast to Amundsen's. The Norwegian expedition left their winter quarters on 19 October 1911,

A camp en route to the South Pole in 1911. Amundsen's strategy relied on his team being small, lightweight, fast and was heavily dependent on dog teams to haul sledges.

Amundsen in Ny-Ålesund, at the far north of Svalbard in 1926, standing next to one of two Dornier flying boats in which he reached a "furthest north" latitude for aviation. He is wearing the sealskin clothing favoured since his time with the Netsilik.

Amundsen leading a party of five men on four sledges with fifty-two dogs. Fifty-seven days later, having covered some 1,720 km (1,070 miles), they became the first men to stand at the South Pole. Amundsen, who had spent his life preparing for the North Pole, would later write of the moment he stood at the opposite end of the planet, "Never has a man achieved a goal so diametrically opposed to his wishes." They made the return to the Bay of Whales in just 39 days and almost immediately departed in the *Fram* for Australia. Scott finally reached the South Pole on 17 January 1912, a full 41 days after Amundsen.

The speed and efficiency of Amundsen's incredible achievement has arguably made it seem almost too easy – but the party did encounter many challenges, several of which could easily have proven fatal. The outcome of the expedition might have been very different.

Amundsen did not believe in luck, famously declaring, "Victory awaits him who has everything in order – luck, people call it. Defeat is certain for him who has neglected to take the necessary precautions in time; this is called bad luck." Even so, a consequence of living on resourcefulness as Amundsen did, is undeniably an increased vulnerability. Amundsen faced many situations during his expeditions that involved taking apparently suicidal risks, yet his confident decisiveness always seemed to win through. During his discovery of the Northwest Passage, his robust wooden sealing ship was blown onto rocks in a severe gale. In danger of being smashed to pieces, Amundsen ignored a fundamental rule of the sea and ordered the sails to be raised into the gale-force winds. He chanced the precious sails being torn to rags and the ship being blown to destruction, but calculated that there was no other choice. Mercifully, his plan worked, and with the sails raised, the winds blew the ship clear of the rocks. The ship, and the expedition, were saved.

In the end, this reliance on his talent for adaptability, which had so often enabled his survival and success, proved fatal for Amundsen. In 1928, he set out in a prototype flying boat to search for Umberto Nobile, a man with whom he had once led an expedition but had since come to despise. Nobile had gone missing with his crew in an airship over the Arctic Ocean. Amundsen's decision to undertake the bold rescue mission in a twin-engine plane with open cockpits over the hazardous and isolated northern seas seems in keeping with his characteristic decisiveness in the face of risk. Yet the exposure, relatively untested capability and hasty departure of the flight is simultaneously at odds with his lifelong insistence on rigorous planning and diligent preparation. In a chilling echo of the fate of his hero, Sir John Franklin, Amundsen, his five companions and the plane all vanished into the Arctic, almost without trace. Amundsen met his end in the north, the region that had been his life's fascination, and we can be certain that he would never have called his death bad luck – but rather, a failure to adapt.

ADOPT THE ADAPTABILITY OF AMUNDSEN

IN HIS EMPHASIS ON ADAPTING TO AN ENVIRONMENT, RATHER THAN SEEKING TO OVERCOME IT THROUGH "BRUTE FORCE", AMUNDSEN REVOLUTIONIZED METHODS OF EXPLORATION. HIS LIFE EXPERIENCES AND EXPEDITIONS ARE A DRAMATIC DEMONSTRATION OF THE ADVANTAGES – AND LIMITATIONS – OF THIS APPROACH, AND PROVIDE INSIGHT THAT HAS RELEVANCE FAR BEYOND EXPLORING.

Amundsen on the ice-laden Ross Sea, Antarctica, in 1911, before setting out on his attempt to become the first man to reach the South Pole.

RESOLUTE FOCUS

From an early age and throughout his life, Amundsen maintained unwavering focus on his ambitions. This clarity of vision enabled him to use his efforts efficiently and to concentrate only on matters that brought him closer to achieving his aims. He was very adept at identifying what was important, even when under pressure and in complex situations. It was this tenacious refusal to be distracted that enabled him to have the flexibility to adapt without losing sight of his goals.

KNOWLEDGE IS KEY

In order to be adaptable, Amundsen believed that it was crucial to have a detailed understanding of any skills or tools on which he might rely as a polar explorer – from navigation at sea or becoming a proficient skier to the Arctic survival techniques of the Netsilik. He invested years of considerable effort to gain the knowledge that allowed him a high degree of self-sufficiency and a resourcefulness that became legendary.

LEARN FROM THE MISTAKES OF OTHERS

Whether by analyzing the experiences of past explorers in written accounts or being a shrewd observer of those he worked with, Amundsen made a habit of distinguishing what had made others successful – and, more particularly, what might have caused them to fail. He was as quick to adopt ideas he thought useful as he was determined to avoid the pitfalls suffered by others.

HUMANITY

NEIL ARMSTRONG

MORE THAN A BILLION PEOPLE LISTENED AS NEIL ARMSTRONG, NEARLY 400,000 KM (250,000 MILES) AWAY, TOLD THEM: "THAT'S ONE SMALL STEP FOR MAN, ONE GIANT LEAP FOR MANKIND." IT WAS A MOMENT THAT BROUGHT ALL OF HUMANITY TOGETHER IN A NEW UNDERSTANDING OF THE NATURE OF OUR EXISTENCE. ARMSTRONG BECAME IMMORTALLY FAMOUS AS THE FIRST MAN TO MAKE THAT HISTORIC STEP ON THE MOON, BUT IT WAS THE HUMANITY AT THE HEART OF EVERYTHING HE DID THAT ESTABLISHED HIS GREATEST LEGACY.

It is hard to imagine how Armstrong might have felt as he returned to Earth as the first human in history to have set foot on another planet. Having blazed through the atmosphere in the Apollo 11 capsule to splash down in the Pacific Ocean west of Hawaii, Armstrong and his crew – Buzz Aldrin and Michael Collins – were collected by a US Navy aircraft carrier before spending three weeks in quarantine together. With the formalities of their return completed, they were launched into what we can suspect must have been a more arduous ritual for these professional astronauts: a relentless schedule of public appearances. From ticker-tape parades and award ceremonies to press interviews and private audiences with key figures over the course of a 38-day world tour to 22 countries, they attracted an unprecedented intensity of attention. Armstrong was a man who had a talent for coolness under pressure, but this new role brought a different kind of pressure he had not experienced before. Not only did he bear the responsibility of representing humankind at the moment it became an interplanetary species, but the world now looked to him for meaning in the new perspective created by his journey. He recalled for the breathlessly waiting billions hanging on his every word the moment he had looked back at his home planet from the Moon: "It suddenly struck me that that tiny pea, pretty and blue, was the Earth. I put up my thumb and shut one eye, and my thumb blotted out the planet Earth. I didn't feel like a giant. I felt very, very small." He was relating a memory, but for the rest of humanity his experiences resonated as profoundly metaphysical. Individuals, groups and whole societies began projecting their own meanings onto not just the lunar mission but Neil Armstrong himself.

By the end of his life, Armstrong was almost universally characterized by the media as having become a recluse – the suggestion being that through privacy and seclusion he sought respite from humanity's overwhelming expectations of him following Apollo 11. Indeed, just three years after his famous steps on the Moon, Armstrong retired from the National Aeronautics and Space Administration (NASA), having never returned to space. Rather than fulfilling what was – even by the 1970s – the common path of celebrity, he sought neither political influence nor great wealth. Nor did he wish to be seen as the champion of any particular cause. Instead, he became a university professor, teaching aerospace engineering, and bought a farm in his native Ohio. He rarely gave interviews and was fiercely protective of both his privacy and the use of his name, image and words. Yet his children are vocal in their repudiation of the characterization of their father as a recluse, remembering him instead as fun, musical and thoughtful. His son Mark recalled in an interview for the BBC in 2019 that Armstrong "wasn't the kind of dad that would tell you what to do all the time. He was more a professor type who would show you different options and urge you to think carefully about your choices and pick the right one, just as he did by example throughout his life."

Armstrong in 1960 standing next to the North American X-15 rocket plane following one of seven missions he flew in the highly experimental aircraft to explore the limits of hypersonic and supersonic flight.

Looking at Armstrong's life and career, it would be entirely understandable to detect a dramatic transformation after his lunar "first" – but instead, what is remarkable is the extraordinary consistency he maintained before and after Apollo 11. He was the same person after stepping on the Moon and becoming the most famous man in the world as he had been before. Tellingly, Armstrong preferred to refer to himself as neither an astronaut nor a pilot but as an engineer. Indeed, it was with the methodical objectivity and calm rationale of an engineer that he successfully navigated the pitfalls of fame. He proved himself as able to maintain admirable composure in the full glare of the

Seated in the back of the lead car, Armstrong (far right), alongside his crewmates Michael Collins and Buzz Aldrin, wave to crowds during a ticker-tape parade along Broadway in New York City to celebrate the Moon landing.

world's attention as he had remained collected in many potentially fatal instances at the controls of an aircraft in his flying career.

Armstrong learned to fly at a young age, earning his licence by the time he turned 16, before he had learned to drive a car. Having accepted a US Navy scholarship, he saw his aeronautical engineering studies at university interrupted in 1949, when he was drafted into the Navy. He was trained as a fighter jet pilot, and at the age of 21 was sent to participate in the war between North and South Korea – in which the United States was supporting the South. He would fly 78 combat missions during the war, but on just his seventh mission, while bombing a bridge in a F9F Panther fighter jet, he hit an anti-aircraft cable, which sliced off almost half of the right wing of his plane. Speeding from 150 m (500 ft) towards the ground while travelling at 350 kts, Armstrong's decision-making was swift and skilful. With great difficulty, he managed to keep his incapacitated jet in the air until he reached a friendly airfield manned by US Marines. There, he successfully ejected from his Panther at lethal speeds, despite his only training in the procedure being a brief talk through the basics by a friend a few weeks beforehand.

After the war and having finished his interrupted studies, Armstrong became a test pilot of experimental high-speed aircraft at Edwards Air Force Base, operated by the National Advisory Committee for Aeronautics (NACA) – which would later become NASA. His job at Edwards was to fly some of the most dangerous aircraft in the world while deliberately pushing the technology to the limits of its capability. It was such hazardous work that the life expectancy of test pilots at Edwards was described as "grim": just a few years before Armstrong took up his position, Edwards had suffered the staggering loss of 62 test pilots, killed over a single period of nine months. During his seven years as a test pilot, Armstrong flew some 200 different types of aircraft, including the North American X-15, which remains the fastest and highest-soaring aircraft ever built. Designed to explore the limits of both supersonic and hypersonic flight, it could reach speeds of 7,274 km/h (4,520 mph) and altitudes of 108 km (354,000 ft). While probing the potential of rocket planes as powerful and unpredictable as the X-15, Armstrong would be faced during the course of his career with numerous terrifying situations, such as exploding engines in mid-flight, unresponsive controls, G-force limiters that didn't activate when relied on, failed landing gear and being skipped off the edge of the atmosphere like a stone on a pond – all of which he consistently confronted with composure and inspired pragmatism.

This extensive exposure to high-pressure danger since his twenties perhaps explains Armstrong's confidence during the last moments of the attempt to land the Lunar Module *Eagle* on the surface of the Moon. Onboard computers were guiding the landing craft toward an area littered with boulders. Realizing it would be a disastrous place to touch down, Armstrong switched off the autopilot and took manual control, piloting the *Eagle* to a better location. He landed with just 30 seconds of fuel to spare. The margins could not have been narrower or the risks more acute, yet his voice was calm and steady as he reported, "Houston. Tranquility base here. The *Eagle* has landed." Armstrong's recorded biometric data later revealed that, despite this apparent sangfroid at the moment of touchdown, his heart was racing at 156 beats a minute. He would later admit, "there were just a thousand things to worry about". Armstrong was as cognizant of what was at stake as anyone – and as human – making his achievement even more impressive. This reassuring humanity became ever more apparent in the years that followed his return to Earth.

Just three days after Armstrong was cleared from quarantine, he was asked to make an address from the main plaza of the United Nations General Assembly building in New York. That same day, he and his crew had already attended parades through Chicago and Manhattan, waving to more than six million well-wishers lining the streets. Later that evening, they were due to be guests of honour at a lavish state banquet in Los Angeles, where they would each be presented with the Congressional Medal of Honor. Regardless of the intensity of the moment, Armstrong used the opportunity – as he would on so many occasions to come – to ensure the wider significance of the Moon landing was not lost amid all the triumph and spectacle. He told the United Nations: "I can tell that you share with us, the hope that we citizens of Earth, who can solve the problems of leaving Earth, can also solve the problems of staying on it."

Significant in unravelling how Armstrong was able to remain so grounded and true to himself, despite all the adulation, is understanding how he perceived his role in Apollo 11. Consistently throughout his time at NASA he was quick to recognize the unseen thousands of engineers, scientists, designers, technicians, administrators and specialists who had all played a part of equal significance – as he saw it – to his own. Armstrong never considered the Moon landing as an achievement that belonged to him. Instead, he knew it to be an "achievement that a third of a million people had been working for a decade to accomplish". Before, during and for many decades after he went to the Moon, Armstrong sent personal letters of gratitude to various teams whose work had been vital to the success that he felt should be claimed by – and credited to – all.

On leaving NASA, Armstrong was selective about the public interviews and appearances he agreed to and guarded about attempts to attach commercial or political associations to his achievements. Yet far from being recalcitrant, he was in private a dedicated communicator. Committed to ensuring humanity was inspired to progress and grow by the accomplishment of landing on the Moon, he personally replied to a staggering proportion of the overwhelming volume of letters

Taken from the Apollo 11 spacecraft in lunar orbit before the landing, a view of Earth at a distance of almost 400,000 km (250,000 miles), partly illuminated by the sun as it rises over the lunar horizon.

Images of Armstrong on the Moon are rare because he was responsible for operating the camera during the landing. However, in this image of Buzz Aldrin, Armstrong appears as a reflection in Aldrin's mirrored visor.

he received over more than four decades. He entered into extensive correspondence with many of those who wrote to him. His generosity of both time and accessibility must have been a considerable burden, but there is no evidence that he ever shied from it. He eventually stopped providing autographs only when he was made aware that they were being sold on the Internet for profit.

It's suggested by those who knew him best that his decision to leave NASA and become a college professor was driven primarily by a desire to teach and pass on knowledge from his extraordinary and unique career. This certainly fits the impression that emerges from his writings, in which Armstrong emphasizes not just the importance of humanity's advancement but also his insistence that progress should be accompanied by a similar focus on the core values we wish to define us as a species. Here lies arguably his greatest legacy. By leaving his footprint on the surface of the Moon, Armstrong inspired generations to think bigger and reach for the stars – quite literally – but it was his quiet adherence to the values he held that encouraged moving forward in the right way, and setting forth to tackle challenges both big and small with unfailing decency and humanity. Of many sagacious quotes the laconic Armstrong left the world, this is perhaps the one that sums up his most precious gift to humankind: "The single observation I would offer for your consideration is that some things are beyond your control. You can lose your health to illness or accident. You can lose your wealth to all manner of unpredictable sources. What are not easily stolen from you without your cooperation are your principles and your values. They are your most important possessions and, if carefully selected and nurtured, will well serve you and your fellow man."

FOSTER HUMANITY LIKE ARMSTRONG

EXPLORERS WHO EXHIBIT COURAGE, DETERMINATION AND AMBITION IN ACHIEVING GREAT THINGS ARE OFTEN SIMULTANEOUSLY IN POSSESSION OF INFLATED EGOS, STUNNING ARROGANCE OR FLINTY RUTHLESSNESS. IT COULD BE ASSUMED THAT THESE CHARACTERISTICS ARE AN INEVITABLE SIDE EFFECT OF SUCCESSFUL EXPLORATION, BUT NEIL ARMSTRONG PROVED THAT THERE IS ANOTHER WAY. HIS INSPIRING HONOURABILITY DID NOT DIMINISH NO MATTER THE SIZE, SCALE OR DANGER OF HIS EPOCHAL ACCOMPLISHMENTS AND INSTEAD PRIORITIZED A SENSE OF HUMANITY.

Armstrong making his historic first step on the Moon in a still taken from footage recorded by a TV camera attached to the outside of the *Eagle*, and watched by 650 million viewers on Earth in almost real time.

HUMILITY

Rather than examining how Armstrong was able to maintain his unshakable modesty despite all the success and admiration, it is perhaps more revealing to recognize that his great successes were enabled largely by his unassuming nature and lack of ego. Humility prevented him from becoming overconfident or complacent and ensured he remained focused and determined.

RECOGNITION

Armstrong went to great lengths to make sure credit was given to the people that he considered had been fundamental to the success of the Apollo 11 mission. He never failed to express – or allowed himself to forget – his gratitude for what he owed to others. Such benevolence was a cornerstone of the humanity he became known for, and ultimately inspired as much respect as either his courage or skill.

THE BIGGER PICTURE

The greater the focus and the more all-consuming the goal, the harder it is to maintain a reasoned perspective. Not only did Apollo 11 give all of humanity an enlarged view of the universe and our place in it, but Armstrong also reinforced the importance of being aware of the long-term impact of exploration rather than the immediate achievement. He reminded us that how we succeed at anything is as meaningful as the success itself.

RISK

AMELIA EARHART

SHE HAD ALREADY FLOWN MOST OF THE WAY AROUND THE WORLD, AND NOW THE ONLY THING THAT LAY BETWEEN AMELIA EARHART AND THE RECORD AS THE FIRST AVIATOR TO CIRCUMNAVIGATE THE GLOBE AT EQUATORIAL LATITUDES WAS A STRETCH OF 11,000 KM (7,000 MILES) ACROSS THE PACIFIC OCEAN. IT WAS A RISKY FLIGHT THAT REQUIRED LANDING ON A TINY ISLAND LOCATED RIGHT AT THE LIMIT OF THE FUEL RANGE OF HER AIRCRAFT, AN ISLAND STARKLY ALONE IN THE VASTNESS OF THE OPEN OCEAN AND BARELY VISIBLE ABOVE THE WAVES.

"Please know I am quite aware of the hazards," Earhart wrote in a letter to her husband on the eve of her departure. Indeed, she had spent several years analyzing and preparing for every aspect of the flight. Her twin-engine aircraft, which had been custom-built for her, was so advanced it was nicknamed the *Flying Laboratory* in reference to its cutting-edge technology. The Lockheed Electra Model 10-E was the first all-metal aircraft and its groundbreaking features, such as retractable landing gear, had been designed with the advent of commercial long-distance flying in mind. Earhart herself had specified the modifications to the brand-new plane. The 10 passenger seats in the cabin had been replaced with fuel tanks that increased the aircraft's range to 6,500 km (4,000 miles), and near the newly constructed navigation station at the back of the plane, an additional window had been added to allow for celestial navigation during flight. As well as a radio transmitter and receiver, the *Flying Laboratory* also had an early radio navigation device: a radio direction finder. This consisted of a loop aerial fixed to the roof of the cockpit, which was hand-cranked to rotate it until a point

Above: Earhart and her Lockheed Vega being swarmed by enthusiastic reporters and well-wishers at Hanworth Air Park in west London, England, after her successful solo transatlantic flight in 1932.

Opposite: Earhart at the controls of the *Flying Laboratory*, the Lockheed Electra 10-E that had been custom built for her attempt to circumnavigate the globe.

of low signal strength was detected in any received transmission. This information would allow the location of the origin of the transmission to be calculated.

Similar care had been taken in the construction of the team that would provide Earhart with vital support during the flight. Her technical advisor was Paul Mantz, a skilled air racing pilot and Hollywood stunt flyer with a reputation for thorough planning. Earhart had sought him out in 1934, three years before her round-the-world flight, to help improve her long-distance flying, and later moved her home from New York to California in order to be nearer to him for training. She also enlisted Captain Harry Manning, who was a ship's captain as well as an aviator, and famous for heroically rescuing 34 crew off a stricken ship in a mid-Atlantic gale. He was an experienced navigator and an expert radio operator with

knowledge of Morse code. Even so, after conducting several cross-country test flights with Earhart and Manning in preparation for the round-the-world journey, Mantz was not entirely satisfied and urged that a second navigator be added to the crew, in particular someone who could focus on the demands of the celestial navigation that would be required. They found Fred Noonan, a man already assured of his place in aviation history for his role in navigating and mapping pioneering commercial airline routes across the Pacific. Like Manning, he was a ship's captain as well as an aviator, but had also spent years as a navigation instructor in the nascent commercial airline industry. Considered one of the best navigators in the world, Noonan was enthusiastic about joining the crew. He told reporters, "Amelia is a grand person for such a trip. She is the only woman flyer I would care to make such an expedition with because, in addition to being a fine companion and pilot, she can take hardship, as well as a man, and work like one."

It was an impressive team, but they were not solely reliant on their own abilities or that of the plane for the successful location of Howland Island, on which everything depended. The United States Coast Guard Cutter *Itasca* had been stationed nearby ready to communicate with Earhart during her flight. The ship also planned to transmit a radio signal which could be used by the plane for direction finding to guide them toward the miniscule island, just 2.5 km (1½ miles) long and 500 m (1,600 ft) wide, and – if needed – the ship's boilers could be used to create a column of smoke for extra visibility. Two other US ships were positioned along the flight route as markers and ordered to burn every light on board. "Howland is such a small spot in the Pacific that every aid to locating it must be available," Earhart warned, but she also faced the risk of their failure pragmatically. Flotation devices and emergency life rafts were loaded for the flight and the crew had rigorously drilled themselves on procedures should they be forced to ditch into the sea.

Despite all the preparation and contingency plans, their greatest asset in facing the dangers of the Pacific was undoubtedly Earhart

The *Flying Laboratory* taking off from Oakland Airport at Alameda, California, in 1937 on its way to Honolulu in Hawaii with Earhart, Noonan, Manning and Mantz on board. The loop aerial of the radio direction finder can be seen fixed above the cockpit.

herself. She was one of the most celebrated pilots of her day, having learned to fly in 1921 at the age of 23. At that time, aviation was in its infancy and was incredibly dangerous. The development of aeroplanes had been greatly accelerated by the First World War, but malfunctions remained frequent – and often fatal – while runways, radio communication and any form of navigation other than by sight had yet to be devised. In the year Earhart began taking flying lessons, all but 9 of the 40 pilots that had been employed by the US government to deliver mail by air were killed by their planes. Earhart, however, was quickly breaking records. Before she had even gained her pilot's licence, she flew to a height of 4,300 m (14,100 ft), higher than any other female aviator at the time. Then, in 1928, she made her first significant mark in the world of aviation by becoming the first woman to fly across the Atlantic. She arrived home from the flight a celebrity, honoured with a ticker-tape parade and later producing a book about her experience. Yet she felt frustrated by the notoriety, adamant that she hadn't earned it. Despite having more than 500 flying hours at the time, she had not been at the controls of the plane at any point during the transatlantic flight, which in her view reduced her role to that of a passenger. "Women must try to do things as men have tried," she asserted. Four years later, that's exactly what she did.

On a bright, cloudless morning in May 1932, Earhart took off from Harbour Grace in Newfoundland in a vibrant red single-engine plane. It was, to the day, the fifth anniversary of the first solo flight across the Atlantic, which had been flown by Charles Lindbergh in 1927 and not repeated since. In a deliberate emulation of Lindbergh's flight, Earhart was attempting the same solo transatlantic journey – a first for a female pilot. The opening challenge of the flight was a malfunction of the altimeter, but Earhart, calmly sipping soup as she watched icebergs from an altitude of 3,660 m (12,000 ft), was confident she could estimate her altitude without the instrument as long as she could still see the ocean. A small fire she spotted later in the exhaust manifold was a more serious problem, but by that time she had no choice but to continue. That's when the weather took a dangerous turn, forcing her to fly in near darkness through volatile thunderclouds. Icy rain froze the controls, sending the plane into a spin, but as it descended, the ice melted and Earhart was able to regain control – only to experience the same cycle of ice and thaw, repeatedly diving in deathly spins towards the ocean. Even when she emerged on the far side of the storms, her ordeal was not over. Due to the spins and climbs in the bad weather, Earhart had used much more fuel than planned, a problem compounded by a leak in the auxiliary fuel tank. It is a credit to her temperament that even after such a mentally and physically tiring flight, she didn't hesitate to abandon her intended course to France. She decided to land in Ireland instead, thereby sacrificing her personal goal of an exact replication of Lindbergh's transatlantic flight, which had ended in Paris. She touched down on fields close to Londonderry in the middle of the afternoon having flown continuously for 14 hours, 56 minutes – all without radio communications or sophisticated navigational equipment. Earhart was used to flying by her wits. She would later write, "The time to worry is three months before a flight. Decide then whether or not the goal is worth the risks involved. If it is, stop worrying. To worry is to add another hazard. It retards reactions, makes one unfit..."

Over the next five years, Earhart would continue breaking records. She became the first woman to fly solo across the continental United States – setting a speed record in the process, which she would better a year later – and then the first person to fly solo from Hawaii to California, from Los Angeles to Mexico City, from the Red Sea to India and more. In 1935, she also flew the first civilian flight to carry a two-way radio, a technological advancement that quickly became the norm. As she gained experience, Earhart also began to win grudging recognition as a competent flier. The *New York Herald Tribune* reported:

"Being men and being engaged in a highly essential phase of the serious business of air transportation, they [airline mechanics] all naturally had preconceived notions about a woman pilot bent on a 'stunt' flight—not very favorable notions, either. It was, undoubtedly, something of a shock to discover that the 'gal,' with whom they had to deal, not only was an exceptionally pleasant human being who 'knew her stuff,' but that she knew exactly what she wanted done and had sense enough to let them alone while they did it. There was an almost audible clatter of chips falling off skeptical masculine shoulders."

Along with her technical expertise, Earhart established a reputation for calmness under pressure and was regularly referred to as "unflappable". It is this quality that is most striking in the transcripts of the radio transmissions made by Earhart as she searched vainly for Howland Island on 3 July 1937: "We must be on you, but cannot see you – but gas is running low. Have been unable to reach you by radio. We are flying at 1,000 feet." The last message was received an hour later and referenced the navigation line that she and Noonan were following, which should have led the plane to Howland Island. "We are on the line 157 337. We will repeat this message. We will repeat this on 6,210 kilocycles. Wait." Earhart and Noonan were declared dead 18 months later. No wreckage or remains have ever been found.

The circumstances of Earhart's disappearance have been minutely examined by a wide range of distinguished experts and enthusiastic amateurs alike in the decades that have followed. It remains unclear why so many of the measures Earhart and her team had put in place failed. Why were they unable to receive radio transmissions from the *Itasca*? Why did the radio direction finding not work? Inevitably, much of this discussion has turned to apportioning blame. Had Noonan correctly taken into account the unusual time zone of Howland Island, or the fact that they had crossed the international dateline – both of which would have made a significant difference to his navigation? Had

Earhart not understood how to work the radio direction finder, which was new technology to her? Manning had left the crew months earlier when the leave of absence from his ship expired, leaving Earhart and Noonan to continue without a skilled radio operator – had this been the critical failing?

Perhaps a more relevant question is whether the tragic outcome of her flight would have made any difference to Earhart's decision to go. It would have been impossible to have removed from the attempt all risk whatsoever and if she had succeeded – which she very nearly did – we would be praising her courage and shrewd management of the risks rather than scrutinizing all her decisions with a perspective biased by hindsight. The history of exploration is littered with stories of greater risks undertaken with far less planning, forethought, experience and professional execution than that of the last flight of Amelia Earhart – but which resulted in success. For explorers of all varieties, risk has always been a very precarious line between adventure and madness.

Above: Earhart and navigator Fred Noonan shortly after successfully landing at Bandoeng in the Dutch East Indies (now Bandung, Indonesia) during their fateful global circumnavigation attempt that would end in their disappearance.

Opposite: The last message from Earhart received by her husband George Palmer Putnam, jotted down as he waited anxiously for news in Oakland, California, on 3 July 1937.

BE AS BOLD IN THE FACE OF RISK AS EARHART

FOR AMELIA EARHART, TAKING RISKS WAS NOT A MATTER OF THRILL-SEEKING. AS A PASSIONATE ADVOCATE FOR GREATER GENDER EQUALITY, WHO BELIEVED WOMEN SHOULD BE ENCOURAGED TO TAKE UP PROFESSIONS TRADITIONALLY RESERVED FOR MEN, SHE SAW THAT HER FLYING AND EXPEDITIONS HAD A WIDER SIGNIFICANCE AND CONSIDERED THIS PURPOSE TO BE WORTH THE PERSONAL RISK INHERENT IN HER ACHIEVEMENTS. NEITHER NAIVE NOR SWAGGERING, SHE ASSESSED RISK CAREFULLY AND HER CALCULATIONS WERE BOTH DELIBERATE AND ANALYTICAL.

Earhart in front of her Kinner Airster biplane in which she flew to 4,300 m (14,000 ft) in 1922, setting an altitude record for women pilots – the first of many aviation records she would hold during her career.

PLANNING AND PREPARATION

Assessing the risk of any venture must be done honestly and systematically. It is through lengthy and thorough planning and preparation that risks are fully identified and steps to reduce those risks can be put in place. "Preparation, I have often said, is rightly two-thirds of any venture," advised Earhart. She offered compelling demonstration of this principle – not only in the considerable development of her team and her plane but also in her willingness to recognize her own limitations and take on extra training to improve her skills where needed.

FEAR OF FAILURE

Earhart perceived failure in a positive way. She saw it not as a lack of success to be avoided but rather as a source of motivation for others to do better and a powerful means by which to encourage progress. Referring to the efforts of women to match the achievements of men, she said, "When they fail, their failure must be but a challenge to others." Thinking this way meant that she was never held back by a fear of failure.

ACCEPTING RESPONSIBILITY

To have confidence in our own appraisal of risk and in assessing what level of risk is acceptable to us, we must be willing to take responsibility for our own actions and decisions – as well as the outcomes that result from them. It is essential to be sure that no undue influence or external pressure is affecting our judgement. Earhart wrote to her husband: "I know that if I fail or if I am lost you will be blamed for allowing me to leave on this trip; the backers of the flight will be blamed and everyone connected with it. But it's my responsibility and mine alone."

LEADERSHIP

SIR ERNEST SHACKLETON

AT 5 P.M. ON 21 NOVEMBER 1915, THE SPLINTERED DECK AND TANGLED RIGGING OF A CRUSHED *ENDURANCE* SANK BENEATH THE GRINDING ICE OF THE FROZEN WEDDELL SEA IN ANTARCTICA. THE IMPERIAL TRANSANTARCTIC EXPEDITION HAD SET OUT WITH THE AMBITIOUS GOAL OF TRAVERSING THE ENTIRE SOUTHERN CONTINENT ON FOOT, BUT IT HAD NOT YET REACHED THE COAST BEFORE DISASTER STRUCK. THE EXPEDITION WAS A CATASTROPHIC FAILURE AND YET THE MAN IN CHARGE, SIR ERNEST SHACKLETON, WOULD EMERGE AS ONE OF THE GREATEST LEADERS OF ALL TIME.

The 28-strong crew of *Endurance* watched her sink from a huddle of makeshift shelters pitched a short distance away on an ice floe. They were now stranded on the thin crust of a frozen ocean above 4,000 m (13,000 ft) of water and nearly 2,000 km (1,200 miles) from the nearest place of safety. They had no communication with the outside world and knew that rescue was impossible. It was late spring in Antarctica and in just a few short months the brief but intense summer would melt and break apart the ice beneath their feet. It is hard to imagine how their situation could have been any more precarious, but Shackleton refused to allow either himself or his men to give in to despair.

During the preceding 10 months, while *Endurance* had been stuck fast in the ice through the polar night of an Antarctic winter, Shackleton had implemented a rigorous routine of activity for his marooned team, including sport and regular entertainments in the evening as well as chores and scientific work. The routine continued now that the company was adrift on the ice, Shackleton recognizing the importance of keeping everyone busy and of maintaining a sense of purpose. He had spent years raising the immense sums needed to finance the expedition, winning investment from private backers as well as the British government and the Royal Geographical Society. As he watched his ship sink, and with it his hard-won expedition, Shackleton must have felt enormous pressure from many sources. There was not only the expectation of his sponsors and a considerable financial burden to contemplate, but also his reputation as an explorer and the anticipation of an entire nation. Just three years previously, Shackleton's great mentor turned rival, Captain Robert Falcon Scott, had died in Antarctica having narrowly been denied the distinction of being the first to the South Pole by Norwegian Explorer Roald Amundsen. The British Empire needed to reassert its prominence in Antarctica, and that duty had fallen to Shackleton. Yet despite all these pressures and his own personal anxieties, Shackleton was resolute in his focus on just one thing above all else – the safety of his men. He abandoned all thoughts of the expedition's objectives, put his crew first and made sure that they knew they were his priority. In return, the crew placed their trust and confidence in him as Shackleton made a series of perilous decisions to get them off the ice alive.

Before *Endurance* finally sank beneath the sea ice, Shackleton had ordered the ship's three lifeboats to be moved onto the ice floes along with materials, equipment, supplies and the expedition's 69 sled dogs. He planned that the expedition would pull the lifeboats across the sea ice toward the distant coast of the Antarctic Peninsula that protrudes northward from the main continent and alongside the Weddell Sea. In preparation for the journey, he instructed that all non-essential items be abandoned – forcing expedition photographer Frank Hurley to choose just 150 photographic plates from the 550 he had taken and to smash the rest (so that the distraught Hurley would not be tempted

The *Endurance* frozen into the pack ice of the Weddell Sea during the Antarctic winter of 1915. Shackleton named the ship in reference to his family motto: *fortitudine vincimus* (by endurance we conquer).

to recover any) and ordering that the ship's cat, a much-loved pet of expedition carpenter Harry McNish, be shot. On the first attempt to haul the lifeboats over the sea ice, the crew managed to progress about 3 km (less than 2 miles) in three days. Trying again in December, they achieved just over 11 km (7 miles) in seven days. Shackleton called a halt and made the decision to change strategy. They would instead make camp and await the breakup of the ice before taking to the water in the lifeboats.

Shackleton had a history of adroitness in making such tough decisions. Three years before Scott and Amundsen embarked on their famous parallel journeys to the South Pole in 1911, Shackleton had come tantalizingly close to claiming the honour himself while leading the British Antarctic Expedition, aboard *Nimrod*. Man-hauling heavily laden sledges with three companions – Frank Wild, Eric Marshall and Jameson Boyd Adams – Shackleton had laboured across more than 1,200 km (750 miles) of treacherous and unexplored glacier ice and snow from the coast of Antarctica to be within 100 nautical miles (180 km, 112 miles) of the South Pole. It was the closest any human had ever been to either pole, and it must have felt to Shackleton and his party that the goal – plus the certain fame, glory and fortune that came with it – was within their grasp. Yet Shackleton made the heartbreaking decision that he and his party would turn back. They had been travelling for 73 days and having already stretched their food supply as far as it would go, they were dangerously exhausted. Concerned that the party would not survive the return journey if they continued to the pole, Shackleton later explained to his wife Emily, "I thought you would prefer a live donkey to a dead lion."

As the crew of *Endurance* sat waiting on the ice, food became scarce and the dogs were shot to ease pressure on supplies. The drift of the pack ice carried the camp far to the north toward the open ocean, but as summer passed and winter approached once again, temperatures began to drop. Shackleton continued to ensure everyone was occupied. While some of the men were experienced polar explorers, others were young scientists, artists or tradesmen with little knowledge of expedition life – one was even a stowaway! – but Shackleton insisted on a unity that was unusual in a strongly class-conscious era. Sailors and scientists took on tasks such as camp chores and scientific data collection equally, and all men shared tents regardless of rank. Even so, not everyone was content with Shackleton's leadership. McNish, the carpenter, was not afraid to voice his dissent and at one point threatened to lead a splinter group that would independently make its own way. Shackleton was quick to exert his authority, and rather than letting the disaffected group leave, he went to great efforts to ensure the crew remained intact. Shackleton's leadership was grounded in a

Second-in-command Frank Wild surveys the wreckage of the *Endurance* as it sinks beneath the ice of the Weddell Sea, having been crushed by the pressure of the surrounding pack.

Shackleton (far left) and the crew of the *Endurance* on a drifting ice floe outside a cluster of makeshift shelters they named "Patience Camp", and which would be their home for more than three months.

strong sense of fairness and unity, but it was not a democracy: the crew called him "Boss" for good reason.

On 8 April, the camp was woken by a loud noise and rose to find that the floe beneath them, which had been their refuge for more than three months, had suddenly split in two. The time had come to take to the boats. The three open wooden rowing boats cautiously nudged a course through dense pack ice, and six days later land was sighted – a foreboding hunk of cliff-edged rock named Elephant Island. It is a desolate place, and grateful as the crew were to finally drag their tiny boats onto a narrow spit of shingle two days later, their situation was in reality little better than it had been on the ice. There was no source of food, barely any shelter and little hope of any passing ships so far from the seal and whale hunting grounds of the Southern Ocean. While two of the lifeboats were used to fashion rudimentary shelters high on the beach, Shackleton ordered the third – named the *James Caird* – to be prepared for a sea voyage. He was planning to make one of the most audacious ocean journeys in the history of modern exploration.

Some 1,300 km (800 miles) to the north of Elephant Island lay South Georgia, where Shackleton hoped to seek help from the perennially operational whaling station in Stromness Bay. The journey would require meticulously precise navigation over the notoriously big and wild seas of the Southern Ocean in the tiny *James Caird,* which was just 7 m (22 ft) in length. If they failed to reach South Georgia or were carried past it by gale and current, there would be no rescue for them or the stranded crew. On deciding who would accompany him on the journey and who would stay behind, Shackleton made perhaps his most astute decision of the entire ordeal. Along with the best navigator and the strongest man among the crew (Frank Worsley and Tom Crean, respectively) he chose John Vincent, Timothy McCarthy and the carpenter McNish – three men who had questioned Shackleton's leadership and who were seen as troublesome. Worried about the negative impact these men might have among the group left on Elephant Island, Shackleton decided it was better that he keep them close by.

The sea voyage in the *James Caird* was every bit as difficult and terrifying as expected. After 16 days of being constantly soaked through with seawater, starved, frozen, half sunk by thick layers of ice forming on the lifeboat's deck and tossed by ferocious gales, the party miraculously landed on the coast of South Georgia. It was a stunning feat of courage, tenacity and skill, which must have made the realization that their torment was not over all the more crushing. They had landed on a coast on the opposite side of the island to their hoped-for destination of Stromness Bay. The rudder of the *James Caird* had broken during the landing, so it was not possible to sail further, and any trek overland would be across unmapped territory – including

Launching the *James Caird* from Elephant Island on 24 April 1916, as Shackleton and five men set out for South Georgia on a perilous 1,300 km (800 mile) sea voyage across the Southern Ocean.

the then unexplored mountains of the Allardyce Range. Deciding there was no alternative, Shackleton set off for Stromness early on the morning of 19 May, accompanied by Worsley and Crean. With neither climbing equipment nor mountain experience, the trio were soon forced to climb to an altitude of 900 m (3,000 ft) before coming across a sheer drop that blocked their path. Exhausted and desperate, they flung themselves down the mountainside, not knowing if it would end in their death. Such was the nature of their journey. After 24 hours of gruelling travel, the three men huddled for a moment behind a large rock. Shackleton instructed Crean and Worsley to sleep but, knowing that if they all slept they might never wake, he roused the men after just five minutes, telling them they had had a full half an hour's rest. Several hours later, they reached the whaling station at Stromness and, for the first time in two years, enjoyed safety.

Two days later, Vincent, McCarthy and McNish were rescued from the landing beach on the opposite coast of South Georgia, but it would not be until August – and after two rescue attempts had failed due to impassable sea ice – that Shackleton finally made it back to Elephant Island with a relief vessel. Remarkably, not a single man had been lost.

The Imperial Trans-Antarctic Expedition had failed to set foot on the Antarctic continent and Shackleton returned home to Britain with neither new territory mapped nor records gained to show for his efforts. Yet he was rightly celebrated as a hero for his remarkable resilience and his leadership, which was distinguished by an inspirational level of care for the men under his command. Sir Raymond Priestley – a man who had served on expeditions in Antarctica under both Scott and Shackleton – famously wrote: "For scientific leadership, give me Scott. For swift and efficient travel, Amundsen. But when you are in a hopeless situation, when there seems no way out, get on your knees and pray for Shackleton."

BE INSPIRED BY THE LEADERSHIP OF SHACKLETON

NOBODY IS BORN A LEADER. LEADERSHIP IS A SKILL THAT MUST BE DEVELOPED LIKE ANY OTHER AND CAN BE MASTERED BY ANYONE WITH THE WILL TO LEARN. SHACKLETON MADE MANY MISTAKES AND WAS CONSTANTLY ANXIOUS OF FAILURE, BUT HE ESTABLISHED AN INTUITIVE LEADERSHIP STYLE THAT WE CAN LEARN FROM TO ENCOURAGE THE BEST OUT OF OURSELVES AND THE PEOPLE AROUND US.

Shackleton (third from left, front row, standing) and the crew of the *Endurance*, all of whom survived being stranded on the deadly sea ice of the Antarctic and returned home safely.

LEADING BY EXAMPLE

In order to inspire the confidence of his men, Shackleton took care to demonstrate in his own behaviour the same qualities he demanded from each member of the expedition. He encouraged hard work and loyalty from the crew through his own willingness to endure hardship and through his palpable commitment to put the well-being of his men first.

MAKING THE TOUGH DECISIONS

Despite his willingness to improvise when the situation demanded it, Shackleton was never reckless in his determination to survive. He was a prudent judge of risk, even if this caused frustration among his men. Crucially, Shackleton was secure enough in his authority to change strategy when he thought it necessary, and was not afraid to abandon plans if they were not working.

UNITY

Shackleton created unity within the crew of *Endurance* by taking the time to build unique relationships with every member of his team and getting to know the strengths and character of each. While maintaining a clear sense of command, Shackleton made himself accessible to the crew so that they felt able to come to him and have their perspective heard. He recognized that a unified team was one in which everyone felt valued.

SELF-DISCIPLINE

NAIN SINGH RAWAT

THERE IS A GOOD REASON YOU MAY NOT KNOW THE NAME NAIN SINGH RAWAT. EVEN AS HIS VAST ACHIEVEMENTS WERE LAUDED BY HIS CONTEMPORARIES IN THE PRESTIGIOUS JOURNALS OF THE ROYAL GEOGRAPHICAL SOCIETY, HE WAS REFERRED TO ONLY AS "THE PUNDIT". HIS REAL IDENTITY REMAINED A SECRET BECAUSE SINGH WAS NOT JUST ONE OF THE MOST IMPORTANT EXPLORERS OF ASIA IN THE NINETEENTH CENTURY, HE WAS ALSO ONE OF THE BRITISH EMPIRE'S MOST SUCCESSFUL SPIES.

At the time of the British Raj in the mid-1800s, China had significant control over the sprawling territory of Tibet, which lay across the Himalayas against India's northern border. Wary of foreign influence spreading into Tibet, China had tightly closed the borders of the country, threatening execution to any foreigner who entered illegally. As a result, an expanse of central Asia larger than modern-day India remained a conspicuous blank on the world map, as perfect a *terra incognita* to the outside world as either of the polar regions. With the aim of collecting intelligence for the British, and at great risk to his life, Singh would expertly fill that blank on the map during a decade of clandestine expeditions. His success, and his survival, was due to his unyielding self-discipline.

Singh was working as a school master in his home village, 3,350 m (11,000 ft) up in the Himalayan border region of modern-day Uttarakhand, when he was recruited by the British military as an explorer-spy. Having been raised in an area that formed a major trading route with Tibet, Singh had travelled back and forth across the border many times, was able to speak the language and was familiar with the complex Tibetan cultural customs. This enabled him to pose as one of the small groups of ethnic traders who were allowed to enter Tibet and, once inside the country, to travel disguised as a Tibetan pilgrim. The Tibetan authorities were well aware that agents from the wider Himalayan region were being used to infiltrate the borders, prompting intense scrutiny of all travellers regardless of their origin. Therefore, despite his background, Singh's cover had to be flawless if he were not to be discovered. He had seen for himself the penalty of detection, witnessing the swift public beheading of an accused man in the Tibetan capital, Lhasa, and recording in his journal that he had felt "alarmed on seeing the summary manner in which treachery in these parts was dealt with".

Singh was tasked with surreptitiously surveying as much as he could of Tibet. As well as finding primary transport routes to Lhasa, he was expected to determine the heights of prominent peaks and the extent of mountain ranges, the course of major rivers, and the location of settlements, in addition to any natural resources such as the salt or gold mines that were rumoured to be abundant. Several technical instruments were needed to carry out the surveying, all of which had to be perfectly disguised or hidden. Singh had secret compartments in his travelling chests for concealing the large sextants required for fixing locations. A clear horizon was needed in order to take measurements with a sextant and when the horizon was not visible, an artificial horizon had to be fashioned by pouring mercury into a dish. Anyone found in possession of mercury would therefore be under immediate suspicion. Singh cleverly hid his mercury inside cowrie shells that were sealed with wax. The ordinary wooden tea bowl he carried, of the type used by most Tibetan travellers, doubled as a dish for the mercury horizon when necessary. A thermometer for measuring the boiling point of water to determine altitude was secreted inside his pilgrim's staff, and Singh carried what looked like a Buddhist prayer wheel but which contained not the expected rolls of prayers but rolls of measurements and notes carefully coded to resemble spiritual invocations.

One of the earliest photographs of Lhasa, taken by the British Mission to Tibet in 1903, some 40 years after Singh had travelled there to calculate its exact location and altitude for the first time.

Taking measurements with a sextant is an intricate task that under any circumstance requires patience and skill. Singh undertook the awkward job despite the hard travelling conditions of the extreme terrain of Tibet. He spent months at altitudes over 4,000 m (13,000 ft), crossing much higher Himalayan passes of up to 6,000 m (19,000 ft) multiple times, all in the depths of winter. On the barren desert plains of the Tibetan Plateau, the distances between staging houses were large, frequently obliging Singh and his companions to camp in rudimentary tents or out in the open, rising before dawn to start the day's march and often continuing late into the night. Firewood was scarce, so for fuel he relied on collecting the plentiful animal dung scattered about. Water was harder to come by, and on one occasion his entire travelling party, having exhausted their supply, went without water for more than 20 hours during a strenuous march across high-altitude plains.

Not only did Singh have to wrestle with testing conditions to make his measurements, he had to do so while remaining undetected by those he was travelling with. To calculate latitude, he needed to use the sextant at night to measure the distance between a star and the horizon, but the unforeseen difficulties were numerous, as described in the official expedition reports: "Reading the sextant at night without exciting remark was by no means easy. At first a common bull's-eye lantern answered capitally, but it was seen and admired by some of the curious officials at the Tadum monastery, and the Pundit, who said he had brought it for sale, was forced to part with it, in order to avoid suspicion." From then on, Singh had no option but to use an oil wick, which was impossible to keep alight in the relentless winds of the Tibetan Plateau. Instead, he taught himself to take the measurements in pitch dark, then carefully set the instruments aside and record the readings the next morning, greatly adding to the effort of an already complex task.

Other survey measurements had to be taken right under the nose of suspicious companions. While on the move, Singh took bearings using a compass hidden in the lid of his prayer wheel and the distance from point to point was measured in paces. It was of critical importance that every single stride Singh took on his gruelling journeys measured exactly 33 in (84 cm) so that he knew he had covered precisely 1 mile (1.6 km) every 2,000 paces. To help him keep count, the Buddhist rosary he carried had been altered to contain 100 individual beads rather than the traditional 108, with each tenth bead being slightly larger. On every hundredth pace, Singh would move a bead and, in that way, keep a tally of the distances travelled. The system relied entirely on his ability to keep his step consistent regardless of terrain or distraction. No matter whether he was travelling uphill over rocky ground, fording streams or descending the steepest mountain pass, his stride had to remain precise throughout.

Singh's task would have been easier if he had travelled alone through the sparsely inhabited country, but the intelligence-gathering aspect of his expeditions demanded that he deliberately attach himself to groups of travelling nomads and trading caravans in order to obtain information. Singh had to develop strategies to be able to focus on the surveying while in company and remain undetected. The expedition reports describe some of his methods:

> It was necessary that the Pundit should be able to take his compass bearings unobserved, and also that, when counting his paces, he should not be interrupted by having to answer questions. The Pundit found the best way of effecting those objects was to march separate with his servant either behind or in front of the rest of the camp. It was of course not always possible to effect this, nor could strangers be altogether avoided. Whenever people did come up to the Pundit, the sight of his prayer-wheel was generally sufficient to prevent them from addressing him. When he saw any one approaching, he at once began to whirl his prayer-wheel round, and as all good Buddhists whilst doing that are supposed to be absorbed in religious contemplation, he was very seldom interrupted.

Over the course of Singh's three major Tibetan survey journeys, each more than 1,900 km (1,200 miles) long over mountainous terrain, he achieved stunning accuracy with an estimated error of just 4 km (2.5 miles). This is where the true magnitude of Singh's achievement becomes most apparent. He didn't just map Tibet, he personally paced every mile with an exactness that required unwavering determination and willpower despite considerable physical and mental hardship. He was able to maintain this rigid self-discipline even in moments of the greatest danger, displaying extraordinary presence of mind. Finding himself on the brink of discovery, which would almost certainly have meant death, Singh calmly sent all of his

Above: A Tibetan Buddhist pilgrim photographed in 1920 holding a prayer wheel of the type Singh used to conceal his survey notes and measurements.

Opposite: The high-altitude plains of the Tibetan Plateau and surrounding mountain ranges, crossed many times by Singh on foot, remain a challenging environment for travel today.

covert information ahead with two of his party while he implemented a genius stroke of subterfuge. Deliberately leaving the majority of his belongings with his landlords in Lhasa to allay their suspicions, he set out with next to nothing toward the monastery that had been his cover story. Doubling back after a few days, and still with next to no supplies or equipment, he started the long journey to the Indian border instead. After many months, he passed into British-controlled territory without being caught. Putting the successful communication of his survey work ahead of his own safety would later be commended by his military handlers as "a piece of true heroism".

Being able to command such self-discipline usually requires strong motivation, but the reason why Singh was so dedicated to survey work for British intelligence is hard to establish. He was a British national, but only due to the fact that the British had claimed and occupied his homeland, which brings into question whether he could have been sufficiently motivated by any sense of patriotic duty. He knew that his identity would remain a secret, so it is doubtful that fame or accolades were a significant incentive. He was promised a generous pension, but his strikingly modest pay for the 13 years spent continually on arduous expeditions or in the training of future explorer-spies makes monetary gain an unconvincing explanation. Singh would not be the first explorer to be drawn purely to the adventure and personal challenge of facing the unknown, as well as the sense of achievement in contributing significant knowledge – but, of course, we can never know for sure. What we do know is that the physical hardship of his expeditions, as well as the mental stress of threatened discovery and the strain of maintaining such rigid self-discipline for so long, took a heavy toll on his health, and particularly his eyesight. In 1875, having completed his sixth prolonged expedition, he was "anxious to retire from active work", physically and mentally exhausted at the age of 45.

Now pensioned, Nain Singh Rawat could finally be unmasked as "The Pundit" and the scale of what he had accomplished staggered the geographical world. The Royal Geographical Society in London

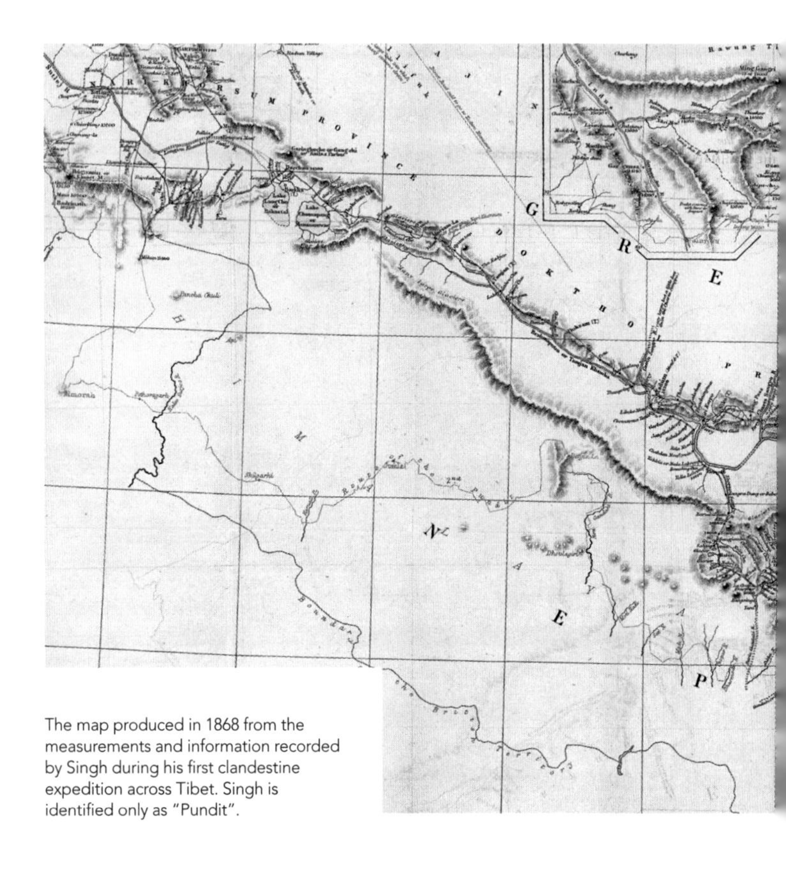

The map produced in 1868 from the measurements and information recorded by Singh during his first clandestine expedition across Tibet. Singh is identified only as "Pundit".

awarded Singh their most prestigious honour, the Victoria Medal, at the end of May 1877, in recognition that his "observations have added a larger amount of important knowledge to the map of Asia than those of any other living man". He had made navigational measurements with a sextant in over 300 locations across Tibet and determined the position of Lhasa for the first time, as well as the height of several major peaks. He made more than 500 observations of the temperature of boiling water, establishing altitudes in so doing, and recorded much additional meteorological data that revealed the climate of large parts

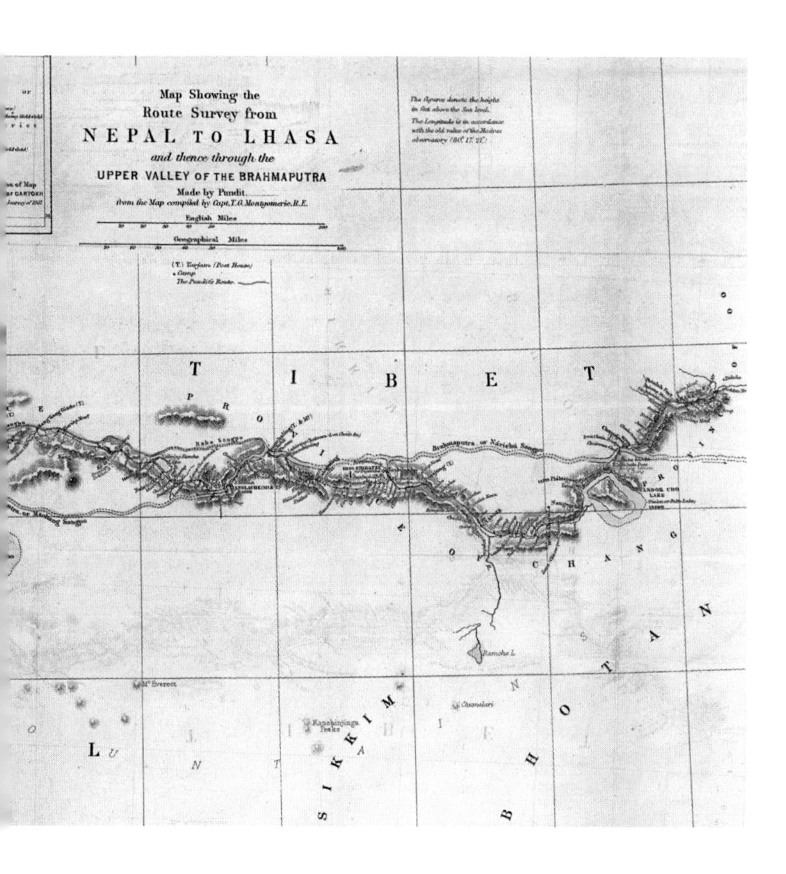

of the region. He had mapped numerous lakes not known before, as well as rivers (including the first charting of the great Brahmaputra River), and even an entire mountain range, which is now named for him. He was able to return with a huge volume of valuable intelligence about the land, its peoples, wildlife and customs – including comprehensive information of Tibetan goldfields. Altogether, he had route surveyed over 6,500 km (4,000 miles) in elaborate detail, the major part of which had not previously been explored. It is a feat perhaps more appropriately appreciated as some 8 million, very disciplined, steps.

HAVE THE SELF-DISCIPLINE OF SINGH

SELF-DISCIPLINE IS VITAL IN OVERCOMING ANY FORM OF ADVERSITY. SINGH WAS ABLE TO MAINTAIN RIGID SELF-DISCIPLINE FOR PROLONGED PERIODS UNDER HUGE MENTAL AND PHYSICAL STRESS. FROM HIM, WE CAN LEARN TECHNIQUES TO IMPROVE OUR CAPACITY FOR STEADFAST SELF-CONTROL WHEN WE NEED IT.

Two local guides and a porter rest briefly on their way over a snow-covered pass through the Himalayas on the Indian side of the Tibetan border. The image dates from around 1880–90, just shortly after Singh was active in these same regions.

ROUTINE

Singh had strict protocols for all his surveying measurements as well as for covert communications and for maintaining his cover stories. Meticulously following prescribed routines makes it easier to stick to a plan and has the simultaneous benefit of increasing efficiency. An action repeated in the same way over and over will quickly become a matter of habit rather than require the effort of self-discipline alone.

GET SUPPORT

Singh was obliged to travel most often with others, but occasionally spent long periods of time alone. This had the benefit of making it easier to hide his survey work, but had disadvantages too. When unaccompanied, the voice of temptation that undermines self-discipline is always present because there is no one there to witness your actions. When surrounded by others, we are more likely to present the best version of ourselves and make larger efforts to implement greater self-control. Self-discipline is easier when we are being observed.

STEP-BY-STEP

Dealing with the levels of uncertainty faced by Singh during his expeditions can be frustrating, disorientating and a drain on both morale and motivation, making it even harder to maintain self-discipline. Being unable to plan effectively, or having plans constantly disrupted, can be mentally exhausting, but vital energy can be saved by avoiding projecting too far into the future. Singh took care to focus only on what was immediate and therefore in his control, even if that was only as far as the next water supply, the next day or the very next step.

COURAGE

THE PICCARDS

AUGUSTE PICCARD APPEARS IN PHOTOGRAPHS AS A VISION OF THE ARCHETYPAL SCIENTIST: TALL AND THIN, WEARING A WHITE LABORATORY COAT, SPHERICAL WIRE SPECTACLES, AND SPORTING UNRULY TUFTS OF HAIR ON EITHER SIDE OF HIS OTHERWISE BALDING HEAD. THERE IS A REASON HIS IMAGE MAY FEEL FAMILIAR TO YOU. THE BELGIAN CARTOONIST HERGE, AUTHOR OF *THE ADVENTURES OF TINTIN*, MODELLED THE CHARACTER PROFESSOR CALCULUS ON AUGUSTE PICCARD AND, IN SO DOING, ESTABLISHED THE STEREOTYPICAL IMAGE OF A SCIENTIST THAT PERSISTS TO THIS DAY.

Auguste Piccard was the archetypal scientist in more than just appearance. Born in Switzerland in 1884, he was at the heart of the scientific crucible that was Europe in the first quarter of the twentieth century. A contemporary of such intellectual revolutionaries as Albert Einstein and Marie Curie, both of whom were personal friends, Auguste studied the magnetization of water, built the most accurate scales and seismographs of the age and discovered Uranium 235 (which he named Actinuran). A professor of physics at the Free University of Brussels, he worked in the tradition of the gentleman polymath, placing an emphasis on the practical rather than the theoretical, allowing his studies to follow his curiosity, custom-building his own equipment and carrying out all his own experiments.

It was while working to validate Einstein's research on relativity that Auguste started investigating cosmic rays – high-energy particles moving at approximately the speed of light. In 1926, he ascended in a balloon to an altitude of 4,500 m (14,764 ft) to observe cosmic rays above the filter of the Earth's lower atmosphere, but it wasn't high enough to allow him to collect the data he needed. At that time, it wasn't unusual to use balloons

Auguste Piccard (third from right) and his engineers standing in front of the spherical aluminium capsule that Piccard would use to make the first manned ascent into the stratosphere.

for scientific experiments (Victor Hess, who discovered cosmic rays, did so from a balloon in 1912), but the stratosphere, that region of the Earth's atmosphere above 10,000 m (32,810 ft), was considered impenetrable. At less than one tenth of the pressure experienced at sea level and at temperatures as low as -60°C (-76°F), the stratosphere was understood to be as decidedly fatal for human beings as outer space. Yet, led by his determination to better observe cosmic rays, Auguste designed and built a small spherical capsule that he calculated would be able to maintain sufficient pressure inside for a human to survive the stratosphere. Needing a mechanism to transport the capsule to stratospheric altitudes, Auguste designed and built a special balloon. Made of rubber, sandwiched between cotton layers that had been soaked in a yellow chemical called chloramine to protect against potential erosion by the solar radiation it would be subjected to in the stratosphere, and filled with hazardously flammable hydrogen gas, the balloon could inflate to a diameter of 30 m (100 ft) but would take its complete shape only at altitude, when it reached lower pressures. On take-off, the balloon took the rather unreassuring form of a withered pear.

On 27 May 1931, in Germany, Auguste and his colleague Paul Kipfer readied for the first attempt to ascend into the stratosphere. With a diameter of just 2 m (6.5 ft), the capsule was large enough for only two passengers alongside the scientific payload. Oxygen was injected into the airtight capsule, allowing the occupants to breathe, and Auguste computed that the sphere's disturbingly thin aluminium shell (just 3.5 mm) nevertheless had a 24-hour endurance – enough for their planned return journey. If anything went wrong, or had been miscalculated, the pair risked dropping to Earth in a fatal free fall, suffocating or being killed by the increased exposure to solar radiation.

Despite meticulous preparation, the journey did not start well. Auguste and Kipfer entered the capsule to make last-minute adjustments to their instruments, then realized, just minutes later, that they were already airborne, the balloon having been launched prematurely. Temperatures inside the capsule quickly started to fluctuate, from sub-zero to over 40°C (104°F), and when the water supply ran out, the crew were forced to drink the condensation that formed on the inside of the sphere. An article published in *Popular Science* shortly after the attempt describes the other dangers of the flight:

During the ascent, the aluminium ball began to leak. They plugged it desperately with Vaseline and cotton waste, stopping the leak. In the first half an hour the balloon shot upward nine miles. Through portholes, the observers saw the Earth through copper-coloured then bluish haze. It seemed a flat disk with upturned edge. At the ten-mile level the sky appeared a deep, dark blue. With observations complete, the observers tried to descend, but couldn't. While their oxygen tanks emptied, they floated aimlessly over Germany, Austria and Italy. Cool evening air contracted the balloon's gas and brought them down on a glacier near Ober-Gurgl, Austria, with one hour's supply of oxygen to spare.

Lasting 17 hours, the flight reached an altitude of 15,780 m (51,771 ft), making the crew the first humans to explore Earth's upper atmosphere, as well as the first to see the profound curvature of the planet. Auguste collected sufficient observations of cosmic rays to be able to validate Einstein's theory of relativity, but in the process had simultaneously invented the stratospheric balloon – still used a century later for upper-atmosphere exploration – and developed technology that would lead to the pressurized cabin used in modern aircraft and enable the high-altitude aviation that has shaped so much of today's world. Auguste continued to improve on the design of his capsule, building versions in which 26 further high-altitude flights were made, achieving a maximum altitude of 23,000 m (75,459 ft). Amid this success, he realized that the technology and principles which allowed him to explore the stratosphere could also allow him to explore another region of the world thought impossible for humans to survive: the deep ocean.

In 1948, off the coast of West Africa, Auguste made the first dive under the ocean in his "underwater balloon", which he called a bathyscaphe. The success of the dive in proving his prototype invention was slightly diminished when the craft became damaged in bad weather. However, Auguste's 26-year-old son, Jacques, was part of the expedition and from his participation would emerge a more significant outcome.

LE PETIT JOURNAL
ILLUSTRÉ
HEBDOMADAIRE · 42e Année
61, rue Lafayette, Paris
7 Juin 1931 - N° 2111
PRIX : 50 CENTIMES
LE MAGNIFIQUE EXPLOIT DU PROFESSEUR PICCARD

Front page of French newspaper, *Le Petit Journal Illustré*, depicting Auguste Piccard landing on the Gurgl Glacier in the Austrian Tyrol after his flight into the stratosphere with Paul Kipfer on 28 May 1931.

Jacques began working with his father on building a new bathyscaphe, which they named *Trieste*. At just over 18 m (59 ft) in length, the *Trieste* comprised a spherical cabin for a crew of two and a cylindrical float filled with gasoline – a substance chosen because it was buoyant but, due to being liquid, was less compressible at pressure than a gas. In 1953, father and son piloted the *Trieste* in the Mediterranean to a depth of 3,150 m (10,334 ft), withstanding pressure more than 300 times greater than at the surface. This was deeper than anyone had ever been before, but Auguste and Jacques were confident the *Trieste* was capable of more.

In January 1960, guided remotely by his father, Jacques began a record-making descent, accompanied by US Navy Lieutenant Don Walsh, into the Pacific to explore the deepest part of the world's oceans, a region known as the Mariana Trench. During the journey to the ocean floor, oxygen was pumped into the cramped cabin of the *Trieste* and carbon dioxide scrubbed from the recirculated air. As the *Trieste* passed a depth of 9,000 m (30,000 ft), the craft trembled when one of two Plexiglass windows, each 15 cm (6 in) thick, cracked. After nearly five hours, the *Trieste* finally touched down at Challenger Deep, the deepest part of the Mariana Trench. The two men had reached a depth of 10,916 m (35,813 ft), which is around 2,000 m (7,000 ft) deeper than Mount Everest is high, and were as stranded from any hope of rescue should anything go wrong as if they had been astronauts landing on the Moon. Spending just 20 minutes at the bottom, the *Trieste* made the return journey to the surface in a little over three hours. The dive was a phenomenal achievement, both in terms of courage and technology – the scale of the feat emphasized by the fact that there would not be another manned dive to Challenger Deep until 52 years later.

For Jacques, the most important accomplishment of the *Trieste* expedition was the observation of life in the deepest parts of the ocean, where it had previously been assumed life of any kind was impossible. The discovery transformed attitudes worldwide toward the ocean, and plans for dumping toxic waste in ocean trenches by the US and other national governments were permanently scrapped. Jacques devoted the rest of his deep-sea explorations to the promotion of the environment. In the tradition of his father, he predominantly worked alone, designing, constructing and piloting submersible craft without the resources of large organizations and yet producing technology of far greater capability than government-funded navies. In 1966, Jacques began working on new medium-depth submersible designs that he termed "mesoscaphes", and in 1969, the mesoscaphe *Ben Franklin*, with an international crew of six NASA and US Navy scientists onboard, submerged 300 m (1,000 ft) into the Atlantic Ocean to explore the Gulf Stream – the first expedition of its kind. This vast ocean current, vital to global climate and ecosystems, carried the *Ben Franklin* north for 2,323 km (1,444 miles), with Jacques and the crew spending a month underwater before emerging on the coast of Maine. Present at the launch of the *Ben Franklin* on its daring and unprecedented journey was Jacques's 11-year-old son, Bertrand.

It is perhaps no surprise that the young Bertrand would grow up to follow the family tradition of exploration. Aquanauts, astronauts, scientists and explorers known by his father were influential fixtures of his childhood and he has recounted his sense of astonishment when realizing in his youth that his belief in exploration as the "only valid way of life" was not shared by everyone. "Exploration frightens those who prefer to take refuge in dogmas, paradigms and assumptions," he writes on his website – and it is clear that Bertrand was not afraid to take on danger in the pursuit of human advancement. Forging parallel careers as a psychiatrist and as a pioneer of ultralight flying, he became European Hang-glider Aerobatics Champion in 1985 while still in his twenties. In 1999, and reminiscent of his grandfather's pioneering flight half a century earlier, Bertrand accomplished the first non-stop circumnavigation of the globe in *Breitling Orbiter 3* – a balloon 55 m (180 ft) tall and fuelled by propane gas with a carbon-fibre and Kevlar gondola suspended beneath. With his British co-pilot Brian Jones, Bertrand launched from Switzerland and landed in Egypt 19 days, 21 hours and 55 minutes later, having travelled a distance of 40,814 km (25,361 miles), climbed to altitudes of 11,737 m (38,507 ft) and achieved speeds of up to 227 km/h (123 knots). At the time, it was the longest distance flight in the history of aviation and remains the longest duration, non-refuelled flight.

The sensation of flying without an engine would cause anxiety in most, but for Bertrand, a skilled and experienced pilot of both hang-gliders and balloons, it was a source of inspiration. In 2003, he initiated *Solar Impulse*, a project that brought together an international and multidisciplinary team of 50 engineers to develop a solar-powered plane capable of flying around the world. The motivation for Bertrand in leading the project was to demonstrate the capability of both renewable energy and human ingenuity in providing a future that protected the environment. A decade later, *Solar Impulse 2* prepared to take off from Abu Dhabi in the United Arab Emirates. The skeletal-looking, single-seater plane, weighing no more than an average pickup truck, had an enormous wingspan of 72 m (236 ft) and was fitted with more than 17,000 photovoltaic cells across the back of the wings, fuselage and tail. Over the next 16 months, *Solar Impulse 2* made a series of 17 flights, piloted in turn by Bertrand and his project partner, (and fellow Swiss) André Borschberg. Flying day and night, using only the power collected from the sun by the photovoltaic cells to soar above oceans, mountain ranges and vast deserts, Bertrand later described the silence of the flight – broken only by the "very soft whistling" of the propellors – as being not unnerving but rather "extremely peaceful". Having travelled 42,438 km (26,369 miles) and spent a total of 558 hours, 7 minutes in the air, *Solar Impulse 2* returned to Abu Dhabi on 23 July 2016, completing the first circumnavigation of the Earth by a piloted, fixed-wing aircraft using only solar power – an achievement destined to be a pivotal moment in aviation as well as the history of human technology and alternative energy.

The courage of the Piccard dynasty lies not just in these feats of exploration but also in the championing of a cause that society had not yet embraced – environmental protection. Unafraid to address the looming and complex challenges facing humanity, in the same way that they have been unafraid to venture into the unknown and hostile regions of the planet, the Piccards have together redefined what is possible when we have vision and courage.

The *Breitling Orbiter 3* above Switzerland shortly after departure on 1 March 1999, with Bertrand Piccard and Brian Jones on board for what would be the first successful circumnavigation of the planet by balloon.

FOLLOW THE COURAGEOUS EXAMPLE OF THE PICCARDS

THERE IS NO EXPLORER GENE TO EXPLAIN THE IMPRESSIVE AND STRIKINGLY PARALLEL ACHIEVEMENTS OF THREE GENERATIONS OF THIS EXTRAORDINARY FAMILY, BUT AUGUSTE, JACQUES AND BERTRAND DO SHARE SOME COMMON TRAITS THAT HAVE ASSISTED THEM IN THEIR COURAGEOUS PIONEERING.

Auguste (sitting, far right) and his son Jacques (standing, centre) on board the *Trieste* in 1953, shortly before making their record-breaking dive in the Mediterranean, to greater depths than anyone before them.

CONFIDENCE

The Piccards each derived courage for their daring expeditions from the high degree of confidence they had in the craft and equipment that carried them. Auguste and Jacques habitually built everything themselves and were intimately familiar with every last detail of their inventions. It was noted of the *Trieste* that not a single dial or switch bore a label because Auguste and Jacques knew the function of every control. Bertrand not only brought together a team of highly skilled experts to create his aircraft, but also oversaw intense testing regimes of his experimental technology. Meticulous preparation is a fundamental source of courage.

PASSION

Auguste Piccard wrote in 1942, "The question now is not so much whether humans can go even further afield and populate other planets, but rather how to organize things so that life on Earth becomes more worthy of living." The sentiment formed a rallying cry that has echoed through three generations of his descendants, each passionate about protecting the environment and seeking a better, more sustainable future for humanity on Earth. Where courage might fail, having a clear and imperative cause is essential.

TRUST

When scientific precision and making the right decisions in the field is a matter not of pride but of survival, choosing the right collaborators and partners is critical in order to build and maintain courage. The Piccards placed trust in their respective co-pilots on each of their daring expeditions and drew confidence from those relationships. Bertrand said of Brian Jones, his co-pilot aboard *Breitling Orbiter 3*, "We took off as friends and landed as brothers."

DIPLOMACY

GERTRUDE BELL

THE PHOTOGRAPH IS A FAMOUS ONE. SITTING ON CAMELS BENEATH THE DISFIGURED GAZE OF THE SPHINX ARE A GROUP OF DELEGATES FROM THE CAIRO CONFERENCE, WHICH HAD CONVENED IN 1921 TO DECIDE THE POST-WAR FATE OF THE NEAR EAST. PICTURED AMONG THEM IS AN ANOMALY: A WOMAN WHO HELD BOTH POLITICAL OFFICE AND MILITARY RANK AT A TIME WHEN IT WAS RARE FOR A WOMAN TO SERVE IN EITHER INSTITUTION, AND WHO WIELDED SIGNIFICANT DIPLOMATIC INFLUENCE OVER A REGION IN WHICH WOMEN WERE FIRMLY DISENFRANCHISED.

Positioned between the unmistakable heft of Winston Churchill, then British Colonial Secretary, and a distracted T. E. Lawrence – Churchill's advisor on the Middle East who was better known as Lawrence of Arabia – is Gertrude Bell. Dressed in a floral hat and fur stole, her appearance belies her position as the senior diplomat who was then managing British control of Mesopotamia as Oriental Secretary. Engaged by British Intelligence, she never wore a uniform but held the formal rank of Major – and was addressed as Major Miss Bell.

Until the First World War, Mesopotamia had been a Turkish province of the once powerful Ottoman Empire. When the Empire collapsed, the province was divided between France and Britain, and it was the administration of this British division that Bell assisted under the military governor. As conference delegates gathered on camels for a photograph at the Sphinx, they had just approved the creation of a new country from the British-controlled territory and agreed the man selected to lead it. It was Bell who had decided the borders of this new country to be called Iraq and it was Bell who had chosen the man sanctioned to be its first king. She had authority over these matters for the same reason that she had more right to pose on camelback than anyone else lined up for the photograph beneath the Sphinx – at that time, she was by far Europe's most accomplished and knowledgeable explorer of the Mesopotamian desert.

A polymath and chronic workaholic, Bell was already a talented linguist, photographer, pioneering mountaineer, diarist, archaeologist and historian when she began making her first desert journeys at the turn of the twentieth century. While still in her twenties, she had spent time with her diplomat uncle in Tehran before travelling the world, from India to the Alps. Returning to the Middle East, she began making short journeys with friends and family from Jerusalem to places of interest, such as the ancient ruins at Petra in today's Jordan. Eventually, she began organizing her own small caravans, striking out as a lone European with a retinue of Arab servants and a dozen camels to explore further from the usual routes and well-known places. In February 1905, she set off from Jerusalem to criss-cross the Levant on a meandering route through what is today Palestine, Jordan and Syria, visiting sites such as Jericho, Damascus, Homs and Aleppo before arriving in Anatolia by April. She was interested in the ancient cities and ruins that were scattered across the region but also the varied tribal and religious cultures that inhabited it. She roamed the haunting landscape of the Jebel Druze to satisfy her curiosity about the secretive Druze population – despite the fact that she lacked the required permission from the ruling Ottoman administration to travel there.

After a short course in survey methods and map projection at the Royal Geographical Society in London – her only formal archaeological training – Bell set off on another great desert journey in 1909, travelling some 2,400 km (1,500 miles) along the Euphrates and Tigris via Baghdad. She had almost instinctively surveyed and photographed ancient sites previously during her travels, but now exploring the archaeological

From left: Winston Churchill, Gertrude Bell and T. E. Lawrence (Lawrence of Arabia) alongside other delegates of the 1921 Cairo Conference, posing for a photograph beneath the Sphinx.

Bell inspecting the ruins of an Arab funerary monument at Duris in Lebanon during an early expedition in June 1900.

character of the lands she passed through became her particular focus. She visited the excavations at Ur before continuing to Ukhaidir, where she directed her own small-scale excavations and conducted the first thorough survey of the site. Already fluent in Turkish, Farsi and Arabic (as well as French, German and Italian), she became additionally proficient in the Bedouin dialect and "accent" of the various tribes she encountered. Benefitting from the custom of hospitality across the Arabian Peninsula, Bell spent long hours as the guest of tribal sheikhs in their desert camps, sharing conversation over Egyptian cigarettes and politely swallowing eyeball delicacies at respectful feasts held in her honour. She spent time in more humble company too, making a habit of joining campfires each evening to listen to tales of Arabic heroes and historic battles in the desert, interspersed with local gossip. Exposed to a broad range of perspectives and knowledge, she began to develop an intimate understanding of the complex social and political life of the desert – as thoroughly appraised of tribal loyalties and the convoluted intrigue of harem matriarchs as she was informed on the best ways to diffuse a blood feud and the looming shifts of influence between prominent families. We can assume that she must have made quite an impression on the sheikhs, especially as she never sought to disguise her foreignness or her attitude of imperial entitlement. As well as travelling with all the supplies you might expect – food, clothing, gifts, medicines, camera and survey equipment – Bell also insisted on having a tent for dining and bathing in addition to a tent for sleeping. Her caravan carried a writing desk and chair, bed and linen, formal dinner set and silverware including candlesticks, the complete works of Shakespeare alongside a carefully selected library and, most elaborate of all, a portable bathtub. Her view on adopting the customs of the Arabs she admired was clear: "The European will be wiser if he doesn't ape their habits; he will meet with far greater respect if he adheres strictly to his own," she wrote.

Not only did her desert journeys teach Bell about the landscape and culture of Arabia, it also taught her the necessary art of desert diplomacy. It was said among locals that "every Arab in the desert fears the other" and indeed, to the Bedouin, looting was a customary activity. The attitude toward foreigners was broadly friendly as long as they were accompanied by a guide who had the right allegiances. These tribal intermediaries for hire were known as *rafiqs*. Bell writes of many tense moments as she travelled between rival tribal territories banking on the abilities of her chosen *rafiq*. Her caravan was regularly intimidated, threatened, occasionally robbed and even shot at. Each time, Bell and her men were saved by a mixture of the right alliances and the right attitude – and being able to change either in an instant according to the situation. The risk lay not just in her contact with the Bedouin but also with the Ottoman authorities that patrolled territory which was still an imperial possession. Fellow explorer Douglas Carruthers remembered Bell returning from the forbidden Jebel Druze in 1905 and after "holding forth in three languages" during breakfast at the hotel, setting off to face the tricky task of explaining herself to the local Ottoman authorities. "By lunch time she was back again, having successfully appeased Turkish apprehensions about an English woman being loose in the Jebel Druze!" he recalled.

In 1913, at the age of 46, Bell would need all her experience, insight and bombastic confidence as she prepared for what would be her last great desert journey. She planned to travel from Damascus to the vast, uncharted and sparsely populated region of Nejd, a desert wilderness of blackened stone crags and bitter winds. At that time, it had been reached by very few Europeans and was consumed by a violent tribal war. The trip was, as expected, a relentless succession of perilous situations that, despite her sanguine narrative in letters and diaries, repeatedly brought even Bell to the brink of despair. Arriving at the ancient citadel of Ha'il, which had been the ultimate objective of her journey, she found herself promptly placed under a genteel house arrest in the company of the palace harem. The amir was away, but after 11 days Bell was led by torchlight to the main palace, not at all certain of her fate. Fortunately, representatives of a tribal group with which she had previously formed ties happened to be visiting the citadel and this, combined with her accurate perception of a palace powerplay at work, appears to have led to her respectful release. The outcome could have been very different.

By this time, the Ottoman authorities had come to regard Bell's travels with suspicion and looked on her as a spy. Whether or not her explorations – and in particular her expedition to Ha'il – had been planned with espionage in mind, there is no doubt that alongside her maps and surveys, Bell passed the social and political information she learned in the desert directly to British Intelligence. In 1915, she was asked to join the Arab Bureau in Cairo to advise the British Foreign Office and allied military intelligence. Ironically, she was to face greater prejudice against her gender in this role than she had ever encountered from Arab sheikhs in the desert. Her military and diplomatic colleagues in Cairo – and later Baghdad – resented her presence but were forced to overlook their opposition because there was simply no one else with the same depth of expertise as Bell. During the First World War, in which Britain and the Ottoman Empire were on opposing sides, every diplomat or intelligence officer entering the Arabian arena received training from Bell. It was her reports about Arab tribal groupings that Lawrence acted on so effectively in 1917 and 1918 in order to persuade the Bedouin to rise in revolt against Ottoman control, precipitating the

One of Bell's own photographs, taken in 1913, of her camp and retinue in the isolated Nefud, a hostile desert that is now part of modern-day Saudi Arabia.

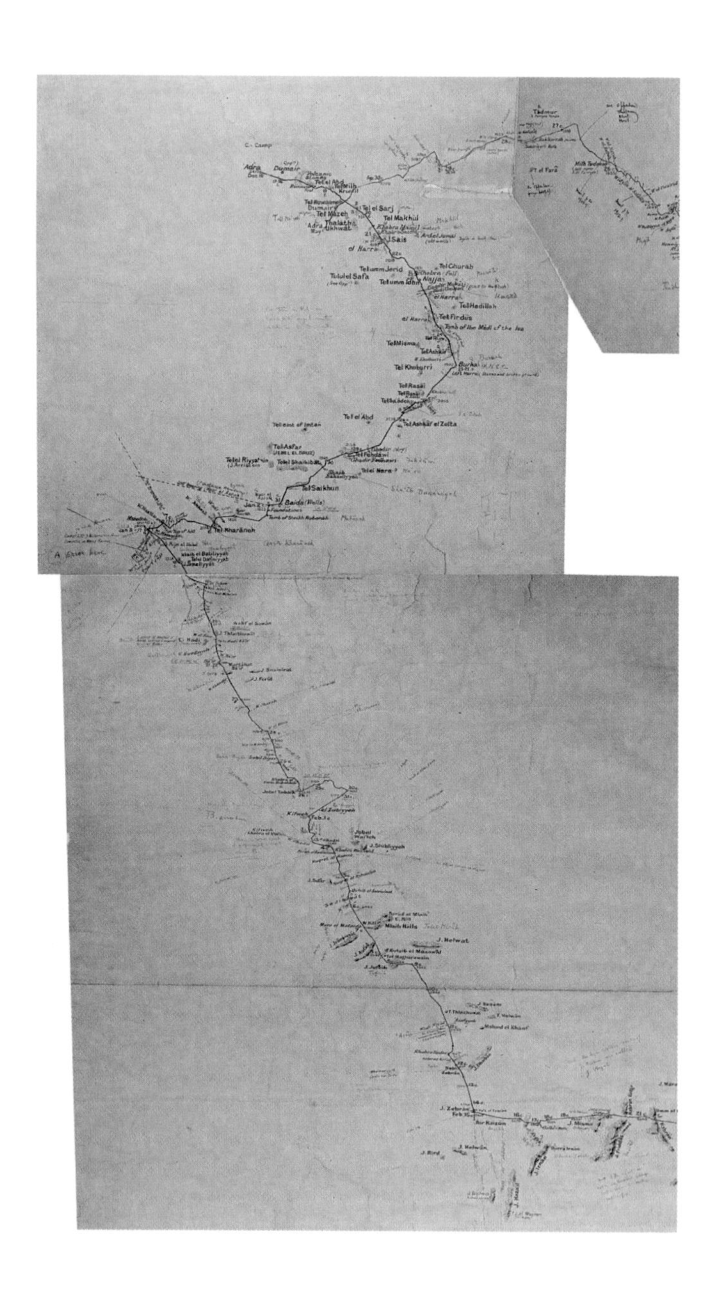
Tadmor
C. Camp
Adra
Dumair
Tel el Abd
Tel Milh
Tel el Sarj
Tel Makhul
Tel Mazeh
Dumair
Thalath
Ukhwat
Adra
J. Sais
el Harra
Tel umm Jerid
Tulul el Safa
Tel umm Idhn
Tel Ghurab
Najjar
el Harra
Tel Hadillah
Tel Firdus
el Harra
Tomb of Ibn Madi of the Isa
Tel Miama
Tel Khuberri
Burka
Tel Rasal
Tel el Abd
Tel Ashhar el Zelta
Tel east of Imtan
Tel Asfar
(JEBEL EL DRUZ)
Tel el Riyyahin
(J. Arrayahin)
Tel el Shaihibat
Tel el Fendawi
Tel el Nara
Tel Saikhun
Jan
Baida (Wells)
Tomb of Sheikh Rubanah
Tel Kharaneh
Shaib Dumaniyeh
el Zebiyyeh
Jebel Wa'leh
Kilweh
J. Shiheiyyeh
Tebuk
Mhaib Haila
J. Helwat
J. Zebran
Bir Kusun
J. Helwan
J. Bird

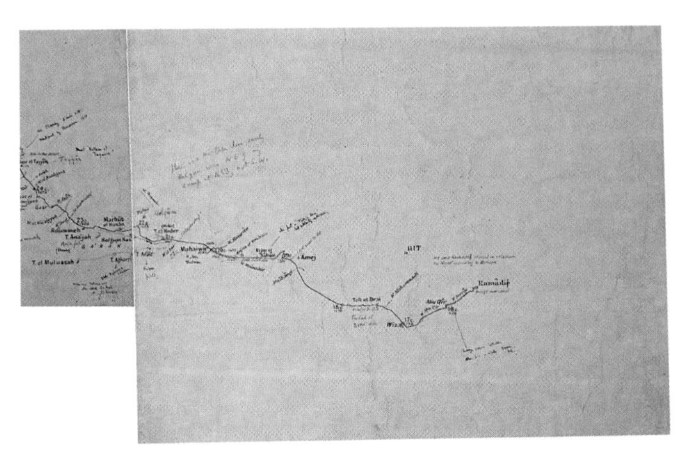

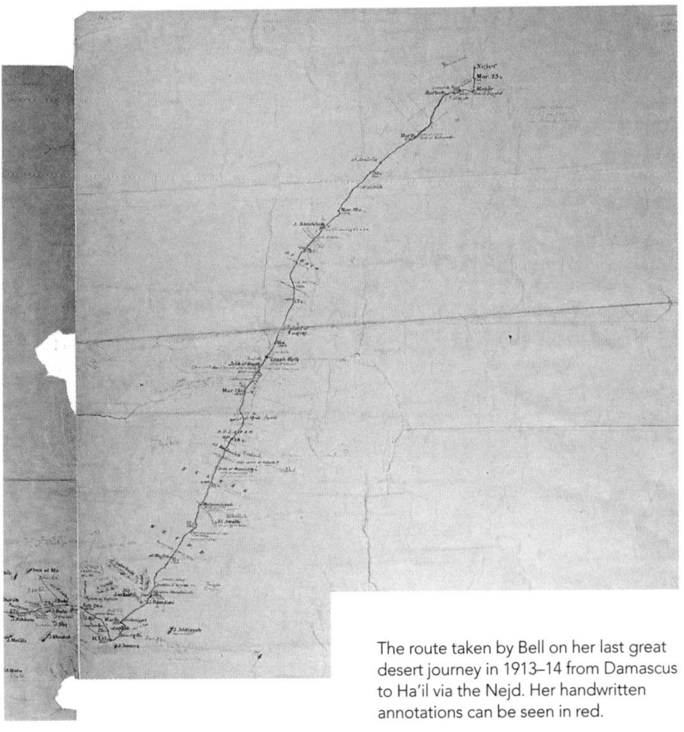

The route taken by Bell on her last great desert journey in 1913–14 from Damascus to Ha'il via the Nejd. Her handwritten annotations can be seen in red.

collapse of the empire. Lawrence motivated Arab forces by promising them self-governance in return, and it was a promise that he and Bell conspired to ensure was kept in the years after the war ended.

Throughout the summer of 1921, following the Cairo Conference, Bell worked tirelessly at the gathering of the inaugural Arab Council (a body she had herself proposed) to convince the leaders of the various desert tribal groups to vote for the new country of Iraq with her choice, Faisal, as king. Many of the sheikhs were men she had met in the desert and with whom she was on familiar terms. Her diplomacy relied heavily on these relationships – a trend emphasized by the fact that the one man she significantly failed to persuade was the one sheikh she had not met on her expeditions. Ibn Saud, head of the powerful Saudi family with influence across the Nejd and much of the Arabian Peninsula, did not see an astute desert explorer when he

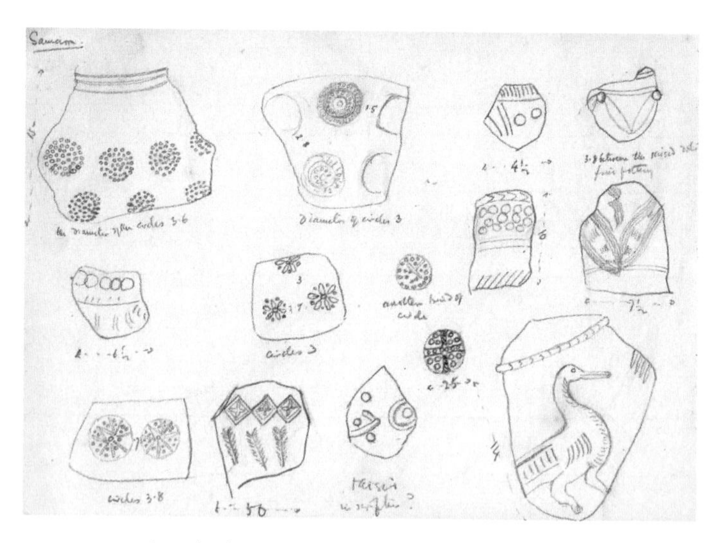

A notebook containing Bell's pencil sketches made in 1909 of pottery fragments she found at Samarra, an ancient archaeological site in what would later become Iraq.

met Bell, but an insult. He was offended that the British authorities appeared to expect him to negotiate with a woman. Arabist St John Philby (who had been trained by Bell) later recorded that Ibn Saud "certainly did not like her ... and many a Najdi audience has been tickled to uproarious merriment by his mimicking of her shrill voice and feminine patter: 'Abdul-Aziz! Abdul-Aziz! Look at this, and what do you think of that?'" The dislike was mutual. Bell wrote of Ibn Saud: "his deliberate movements, his slow sweet smile and the contemplative glance of his heavy-lidded eyes ... do not accord with Western conception of a vigorous personality ..." Yet, she recognized his significance as a powerful force in Arabia at that time, an assessment that others in government failed to appreciate – to their cost. Ibn Saud would eventually form the Kingdom of Saudi Arabia.

When Iraq was duly established and Faisal made its king, Bell made Baghdad her permanent home and worked to ensure the success of the new country. From arranging the daily appointments of the king to writing the constitution, she played an extraordinary role in the shaping of the new nation – but even she was aware of its faults. "There's no getting out of the conclusion that we have made an immense failure here," she wrote in 1920. Many of the devastating challenges experienced by Iraq and the wider Middle East in the twenty-first century have clear roots in the decisions made by Bell and her European colleagues a century ago. Yet, her writings – particularly her legendary dispatches for the inter-intelligence *Arab Bulletin* – are still valued by contemporary diplomats for their insight and acuity. In the years since Bell's untimely death in 1926, a central tenet of Middle Eastern diplomacy has become the development of a greater sensitivity and understanding of the region's history, societal structure and internal politics. We can judge that Bell, whose own diplomacy was grounded in a genuine and profound appreciation of the cultures of Mesopotamia, would have thoroughly approved.

MASTER DIPLOMACY LIKE BELL

YEARS OF EPIC EXPLORATIONS ACROSS THE MIDDLE EAST AND ARABIAN PENINSULA TAUGHT BELL THE SKILLS OF DIPLOMACY - LARGELY OUT OF NECESSITY. HER ABILITY TO DEAL WITH PEOPLE IN A SENSITIVE AND RESPECTFUL WAY ENABLED HER TO GAIN REMARKABLE ACCESS TO ANCIENT ARCHAEOLOGICAL SITES AND SECRETIVE CULTURES, BUT ALSO FREQUENTLY SECURED HER SAFETY. THERE IS MUCH THAT CAN BE LEARNED FROM HER METHODS.

The fateful meeting of Ibn Saud (far left), who would become the founding ruler of the new and powerful state of Saudi Arabia, and Gertrude Bell (third from left) in her official role as Oriental Secretary.

RESEARCH

When faced with difficult situations, Bell drew on her extensive knowledge and understanding of the cultures, tribes and individuals she encountered on her journeys. She gained this resource by dedicating time and effort to cultivating relationships as well as gathering information from a range of sources. Intelligence is the primary tool for effective diplomacy, and it can be seen that Bell's diplomatic failures were in situations where she lacked her usual first-hand insight.

BE AUTHENTIC

Successful diplomacy is built on trust, and it is hard to achieve that without being genuine. Bell never tried to disguise who she was and was unfailingly candid in her opinions. Her insistence on carrying the trappings of her class back in Britain into the desert may appear extravagant, but it signalled to those she met from outside her own culture that she was someone of status, and she was received by them accordingly. Her authenticity garnered respect, even from those who opposed or disliked her.

KEEP PROMISES

Bell associated with many people she knew to be unscrupulous, and yet comprehended the importance of being known as a person who kept her promises. Both Bell and Lawrence understood that the perception of honour was particularly important in the desert societies of Arabia. They both worked hard to ensure the pledges made to Arab wartime allies were honoured, and although they were ultimately unsuccessful in that aim, Bell accurately recognized the consequences that the resulting lack of credibility would eventually engender.

CREATIVITY

AGATHA CHRISTIE

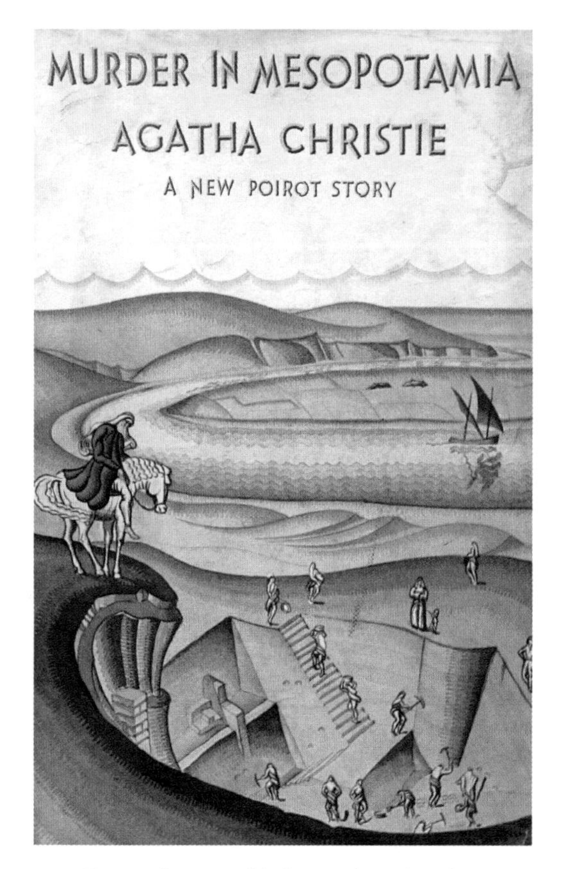

The cover illustration of the first UK edition of *Murder in Mesopotamia* which was published in 1936 and set on an archaeological dig in Iraq.

One of Christie's own photographs of the excavations at Nimrud, the ancient military capital of Assyria, where she completed fieldwork almost every season for a decade alongside her husband Max Mallowan (left foreground, back to camera).

her ingenious plots, her prose often becoming infused with the colour of her immediate surroundings. In particular, *Murder in Mesopotamia* (1936) is set on an archaeological dig in northern Iraq, clearly inspired by her beloved ancient city of Ur, and with characters based on people Christie knew from her time there. Other examples include *They Came to Baghdad* (1951), which capitalizes on Christie's familiarity with the Iraqi city; *Appointment with Death* (1938), which is located among the ruins of Petra; *Akhnaton* (1937), a play set in ancient Egypt; and, most famously of all, *Murder on the Orient Express* (1934), which takes place on the eponymous train. In *Death on the Nile* (1937), a novel set against the backdrop of several ancient Egyptian sites, Christie has her hero, the Belgian detective Hercule Poirot, describe the creative process of her thrilling fiction using the metaphor of archaeology: "In the course of an excavation, when something comes up out of the ground, everything is cleared away very carefully all around it. You take away the loose earth, and you scrape here and there with a knife until finally your object is there, all alone, ready to be drawn and photographed with no extraneous matter confusing it."[2]

Archaeology richly influenced Christie's literary creativity but, in turn, what Mallowan once described as his wife's "peculiar imagination" inspired her uniquely creative approach to archaeology. From 1949 until 1958, Mallowan directed excavations at Nimrud, an ancient Assyrian city situated just south of modern-day Mosul in Iraq. Several thousand rare ivory carvings were uncovered at the site, the largest and most complete collection of ancient ivories ever found. Many of them were discovered submerged at the bottom of a 26-m (85-ft) well, where they had probably been dumped during the sack of the city in the seventh century BCE. It was Christie who was responsible for nursing these rare and important 3,000-year-old figures back to life. Experimenting with different methods of cleaning the fragile pieces, she pioneered a brilliant technique that involved gently applying her own diluted face cream with orange sticks and knitting needles. It is not a method that would be used by conservationists today, but it did result in some of the best-preserved historic ivory carvings in the world. In 2011 the keeper of the Middle East Collections at the British Museum, where many of the Nimrud ivories are now housed, told journalists that they likely exist today solely due to Christie's efforts.

In her photography too, Christie was conspicuous in her creative use of new techniques. During each season of excavation, the job of developing and enlarging a considerable volume of photographs in the field was a difficult one. Christie called on her chemistry background to set up dark rooms in whatever space was available, describing the frequent danger of asphyxiation in the cramped and unventilated makeshift labs. Finding clean water to use for the process was also a challenge, and Christie had to filter water from the nearby Tigris River. Meticulous care was required to prevent desert sand penetrating water, chemicals and equipment disastrously, and even when Christie took to working in the very early morning to avoid the worst of the desert temperatures, the overwhelming heat quickly warmed both water and the lab so as to be unworkable. Given such testing conditions, her consistent and highly competent output was a considerable achievement, especially when she began to take the photographs in addition to developing them.

After a number of years in charge of photography at sites in Syria, Christie enrolled at the Reinhard School of Commercial Photography in London in 1937, keen to improve her skill. There she was exposed to new ideas in composition and perspective as well as the latest technology. Christie embraced it all enthusiastically and returned to her photography work in the field excited to experiment. She began to take pictures not just of the objects and structures unearthed, but also of the digs themselves. In a forerunner of a methodology now common in professional archaeology, she deliberately composed

One of the 3,000-year-old Nimrud ivories discovered at the bottom of a well at the excavation site in 1951 – many of which were initially cleaned and conserved by Christie using her creative techniques.

her photographs to record not just the excavation site and finds but also the people and cultures that placed the archaeology in context. She experimented in film as well. The first of the two films she made was shot in Syria in 1938 and is notable for its early use of colour. The second features Nimrud in Iraq, filmed in 1952 and 1957. Like her photography, the film is significant for its record of everyday life at the excavation, offering a valuable glimpse of a world that would otherwise be undocumented and lost to memory. The resulting visual archive of excavation sites is still used by contemporary researchers today and it is unlikely Christie could ever have envisaged how highly it is now valued for its ethnographic as well as archaeological insight.

The legacy of Agatha Christie's creativity is colossal. In her 86 years she published 66 novels, with a further six under the pseudonym Mary Westmacott, as well as 150 short stories and two memoirs, and became the world's bestselling author, exceeded only by Shakespeare and the Bible. In addition, she wrote 18 plays, including *The Mousetrap*, which remains the longest-running play of all time, and is the only female playwright ever to have three successful plays showing simultaneously in London's West End. As an explorer of ancient civilizations, she similarly had an indelible impact. Long after she left Iraq, Christie's spirit inhabited the British School of Archaeology in Baghdad, which had become her second home while Mallowan was director from 1947 to 1961. Her influence endures in the organizations that have been its inheritors just as it does in the gleaming ivory treasures within the British Museum or the black and white archival photographs of Iraq and Syria in the mid-twentieth century, worlds now physically and culturally all but obliterated by recent wars. Thanks to the nimble creativity of her mind, the name Agatha Christie is associated with archaeological conservation in addition to her murderous plots, and is as deeply embedded in the story of historic antiquities as it is indelibly inscribed into the history of detective literature.

HARNESS THE CREATIVITY OF CHRISTIE

THE CREATIVITY OF AGATHA CHRISTIE MANIFESTED ITSELF AS A REMARKABLE WILLINGNESS TO EXPLORE ALL THE POSSIBILITIES OF A CIRCUMSTANCE. IT WAS A TRAIT DESCRIBED BY THOSE WHO KNEW HER AS AN INFECTIOUS ZEST FOR LIFE. THIS APTITUDE FOR LIVING CREATIVELY CAN BRING HUGE PERSONAL AND PROFESSIONAL REWARD – PLUS, AS IN CHRISTIE'S CASE, CONSIDERABLE HAPPINESS.

Christie pictured in 1946 correcting proofs of her latest literary work at Greenway, the family home in Devon that she shared joyfully with Mallowan.

EMBRACE EXPERIMENTATION

To experiment is to be a pioneer, something that came naturally to Christie. While barely out of her teens and still living in Torquay, she took up the modern sport of rollerskating, and in the 1920s, while travelling around the world as part of a trade delegation with her first husband, she learned how to surf in South Africa and became one of the first Britons to master standing on a surfboard. She is proof that being open to experimentation leads to a life full of discovery.

TECHNOLOGY

Christie's work in photography and film is an example of how technology, used with care and originality, can enhance and inspire creativity to great effect. Christie was unafraid to learn new skills, even relatively late in her life and career, and made a point of investing time and effort to ensure that she was equipped with the expertise to make the most of the latest technological knowledge.

FIND JOY

It is impossible to become familiar with the life of Agatha Christie without being struck by the sense of fun she derived from everything she did. It appears she never allowed the basic conditions of life on excavation sites, the setbacks in her personal life, or the pressures of her writing career to blind her to the wonder in the world. Her creativity flowed from her joy of living, and offers a beautiful reminder to seek enjoyment in every opportunity.

DETERMINATION

JUNKO TABEI

AT 12.30 P.M. ON 16 MAY 1975 JUNKO TABEI AND HER SHERPA GUIDE, ANG TSHERING, STOOD ON THE SUMMIT OF MOUNT EVEREST. TABEI WAS THE FIRST WOMAN OR – AS SHE HERSELF ALWAYS INSISTED – THE 36TH PERSON TO REACH THE SUMMIT OF THE WORLD. HER DETERMINATION IN REACHING THE TOP WAS REMARKABLE, ESPECIALLY AS JUST 12 DAYS EARLIER SHE HAD BEEN ENTOMBED IN A CATASTROPHIC AVALANCHE THAT LEFT HER UNABLE TO WALK.

The incident should have ended the expedition. At 12.30 a.m. on 4 May, a huge section of snow and ice came loose from the flank of Mount Nuptse and fell 300 m (1,000 ft) onto the Western Cwm glacier, where a cluster of small tents had been pitched in what was supposed to be a relatively safe location. Sleeping inside the flimsy shelters were Tabei and six other members of the Japanese Women's Everest Expedition, accompanied by a film crew and several Sherpas. The expedition had already been on Everest for six weeks, climbing in the "siege" style that was then standard on high-altitude peaks. The method involved establishing multiple camps along the chosen route to the summit and using large numbers of climbers and Sherpas to ferry back-breaking volumes of equipment and supplies between them. The expedition would slowly advance up the mountain until a camp was established close enough to the summit to allow one or two members of the team to make a push for the top. At 6,500 m (21,326 ft), in what was known as Camp 2, and having already fixed the route and supplies up to Camp 5, the Japanese team was closing in on its goal and expected to make a final bid for the summit within as little as a week.

Tabei was the team's climbing lead. At 35 years old and at just 1.45 m (4 ft 9 in) tall, she was described by one historian of Everest

Tabei climbing Communism Peak (7,495 m/24, 590 ft) in 1985. It was then the highest mountain in the Soviet Union. Today it is in Tajikistan and renamed Ismoil Somoni.

mountaineers as "not a great climber in terms of expedition climbing abilities" but, significantly, "a determined one". This quiet strength and unflappable determination had already become Tabei's hallmark. When the avalanche struck, she was sharing a sleeping bag with a teammate, the consequence of a rare logistical mix-up. Giant blocks of ice and rubble bulldozed most of the camp 10 m (33 ft) downslope, sending tents, people and equipment violently tumbling and burying them under tonnes of debris. Tabei, smothered by the bodies of her tent companions, would later recall the moment of realization that she was fatally pinned, unable to move or breathe; as she lost consciousness, she was thinking of her three-year-old daughter. Minutes later she was yanked free by Sherpas who had miraculously been missed by the fall. Tabei survived, but she was seriously injured, both by the avalanche and by the brutal rescue that had saved her life. She was unable to bear her own weight, the slightest movement sending shooting pains from ankle to hip; she felt acute chest pain with each breath; and her lower back was severely bruised, leaving her in further agonies.

With Tabei severely injured, and six more of the expedition's 15 climbers similarly hurt and badly traumatized, their tents wrecked and technical equipment lost (including the vital oxygen bottles needed to survive in the thin air of the extreme altitude), and with even the ostensibly invincible Sherpas seriously cut, bruised and shaken, it was clear to everyone that the expedition was over and that the focus should now be a safe evacuation from the mountain. Clear to everyone but Tabei. "As soon as I knew everyone was alive," Tabei said later, "I was determined to continue."

From her incapacitated state in a makeshift emergency shelter, Tabei saved the expedition. Even when the expedition leader and doctor together insisted from Base Camp that those at Camp 2 descend to a safer altitude, Tabei refused. "I knew I had to remain positive and keep Everest reachable in segments," she recollected.

Crucially, she had earned the support of her guide and head Sherpa, Ang Tshering, who declared that if Tabei was staying, the Sherpas would stay too. Slowly, her conviction of spirit persuaded the rest of the expedition that they still had a chance to achieve the summit of Everest. They sewed up damaged tents, fished equipment out of avalanche debris, made repairs with duct tape and three days later, when Tabei was able to take her first unsupported steps, they began to make plans for a summit attempt. Tabei was to tackle the same south-west ridge route used by Edmund Hillary and Tenzing Norgay, famously the first climbers to reach the summit 22 years earlier. Tabei said of the climb, "It was so difficult but I knew it wouldn't last forever. If I reached the top it would be over. I thought, 'If I don't give my all right now, when will I?'"

Tabei (second from right, standing) during an expedition with a team of four Estonian and two Japanese climbers.

For Tabei, continuing when others might have quit had already become a speciality. On every expedition, she'd had to fight, both on and off the mountain, for the right to climb. Japanese society at that time expected women to stay at home, especially when they were a wife and mother. "Some of the men wouldn't climb with me, but a few older ones were more supportive," Tabei remembered of the climbing club where she started. "Some thought I was there to meet men, but I was only interested in climbing." Women were unlikely to be offered places on the big climbing expeditions Tabei craved, so she helped to establish Japan's first all-female climbing club which, in 1970, organized its own Himalayan expedition – to Annapurna III. At 7,555 m (24,786 ft), the peak had been climbed only once before, via a route on the north side. The Japanese team aimed to climb a new route on the south side. In the same year that the famous British Annapurna South Face Expedition led by Sir Chris Bonington was on the mountain (albeit climbing a different route), the Japanese arrived to find that heavy snow prevented the expedition's substantial team of Sherpas and porters reaching Base Camp at the foot of the mountain. Undeterred, the women persevered, transporting much of the cargo themselves and beginning their climb from a much lower altitude than planned. The expedition was persistently plagued by deep team dissension and continued weather difficulties, but Tabei nevertheless overcame every obstacle to reach the summit of Annapurna III on 19 May 1970, along with a fellow expedition climber and two Sherpa guides.

It was the success on Annapurna that inspired the Japanese Women's Everest Expedition, which would take place five years later, but as Tabei and her fellow climbers set about sourcing the necessary funding they needed to make the Everest expedition happen, they found support stymied by entrenched gender discrimination in Japan. "Newspaper articles liked mocking us," Tabei remembered. "They

Tabei on the summit of Mount Everest at 12.30 pm on 16 May 1975.

would use the picture of us applying a lip balm and say 'even in the mountain, they don't skip wearing the make-up.' For a lot of people, it was a joke. They didn't think we would make it."

Lacking sufficient sponsorship, the team scraped funds together themselves by taking on extra work, sacrificing savings and family inheritances. To save money, they made their own equipment. Recycled car covers were used to make waterproof over-gloves, old curtains became climbing trousers, and down was purchased cheaply from China to stuff home-made sleeping bags. Explaining her determination to get to Everest, Tabei revealed, "There was never a question in my mind that I wanted to climb that mountain, no matter what other people said."

Given the prejudice Tabei experienced, it would be easy to assume that a need to prove herself and the capability of her gender might have formed the source of her determination. But if this played a part in her thinking, Tabei never said so. Instead, she was quite clear that it was the mountains themselves that were her biggest motivation. "It is because I love mountains. I love to go wherever I've never been before," she explained. "Climbing the mountain is its own reward." For Tabei, the experience of mountaineering and being in the mountain environment was a joy in itself, even when dangerous and uncomfortable. She described feeling exhilarated to "know what cold and scared really felt like".

Certainly, Tabei had been fully exposed to the stark reality of danger in the mountains. Not only had she narrowly escaped fatality on Everest but she would also experience a further catastrophic avalanche a decade later on Jengish Chokusu, the highest mountain in the remote Tian Shan range on the border between China and Kyrgyzstan. She lost many friends to the mountains too, including her early climbing partner Rumie Saso, a loss that almost crushed Tabei's love of mountaineering with grief. These experiences seem to have given Tabei an acute consciousness of our limited lifespan and the importance of making the most of our opportunity to see the world. "Japan's lifestyle is scary, it's so meaningless," she once told a reporter. "What's the use of building an expensive house if you never really live? If when you are old, you have nothing to look back on?"

Having successfully scaled Everest, Tabei no longer needed to prove herself, but she continued to climb ambitious peaks that challenged her abilities, her determination driven by the purely personal goal of relishing the experience. She summited all five mountains over 7,000 m (23,000 ft) in the Pamir and Tian Shan mountain ranges of Central Asia and several mountains over 5,000 m (16,400 ft) in South America. In 1992, she became the first woman to climb the highest peak on each continent, a feat known as the Seven Summits, and then set herself the goal of climbing the highest peak in each of the 192 countries recognized by the United Nations. "I don't know how or why I'm going to die," she once said. "But I'll look back and think, 'I had an amazing life.' As long as my body is able, I'll keep doing what I love."

She was true to her word. At the age of 77 and having been diagnosed with terminal peritoneal cancer, Tabei insisted on being allowed to leave her hospital bed for one last climb and won a special dispensation from her doctors. She joined what had become an annual expedition with youth from her hometown of Fukushima to the summit of Mount Fuji at 3,776 m (12,389 ft). Despite her illness and through sheer might of willpower, Tabei made it to 3,100 m (10,170 ft) before being forced to descend. She died three months later, having climbed an impressive 76 of the 192 peaks that had been her ambition.

Determined to the end, shortly before her death Junko Tabei was asked in an interview what advice she would give to her younger self. "Do not give up," she replied. "Keep on your quest."

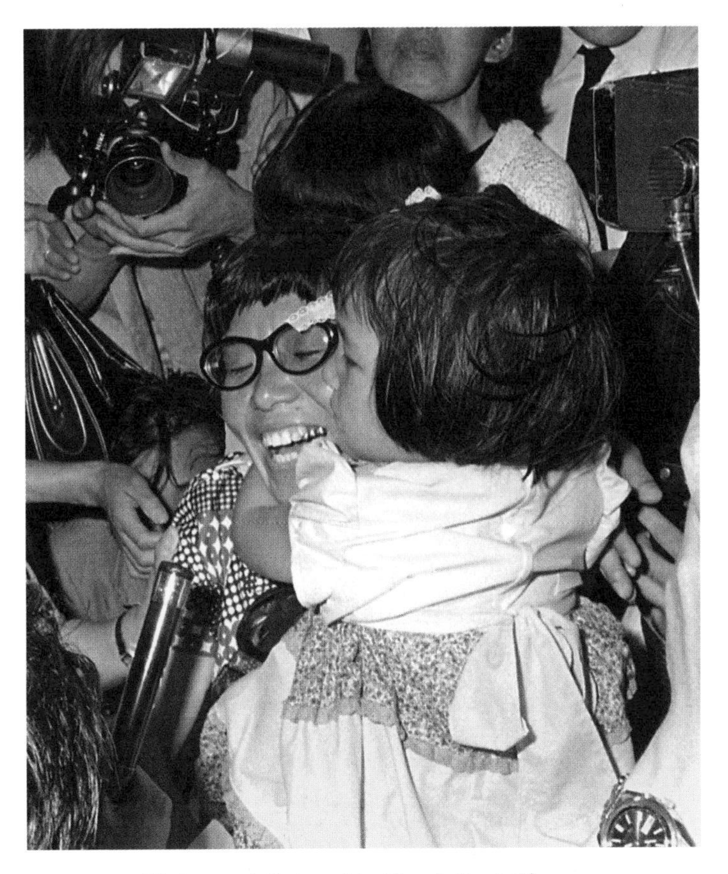

Tabei surrounded by journalists at Haneda Airport, Tokyo, on her return from the successful climb of Everest – and greeted with a tight hug by her three-year-old daughter, Noriko.

BECOME AS DETERMINED AS TABEI

TO FUEL A LIFETIME OF DETERMINED EXPLORATION, TABEI DREW ON HER LOVE OF THE MOUNTAIN ENVIRONMENT AND DESIRE TO FILL HER LIFE WITH AS MANY EXPERIENCES OF NEW PLACES AS POSSIBLE. INSPIRATION CAN VARY WIDELY FROM PERSON TO PERSON AND WE EACH HAVE TO DISCOVER FOR OURSELVES WHAT IT IS THAT WILL FORM OUR OWN UNIQUE SOURCE OF DETERMINATION.

Tabei on Everest in 1975 with lead Sherpa, Ang Tshering. Together, they determined the fate of the Japanese Women's Everest Expedition.

INVEST TIME

Tabei's root motivations were not what might have been assumed, and our own might be equally difficult to identify, requiring a deliberate investment of time and effort to uncover. It can be difficult to unpick precise motivation from the wealth of contributing factors and influences that affect our decisions, but being as precise as possible about what galvanizes us to give our best can be a powerful tool to have ready in those moments when we need to call on some extra inspiration.

BE THE NAIL

There is a Japanese proverb that says, "The nail that sticks out will be hammered down." Tabei often spoke about her willingness to be the "nail that sticks out" and encouraged others to be similarly resolute in the pursuit of their ambitions and unafraid of forging their own paths. By accepting, and even expecting, that she would be challenged and confronted by negative attitudes, Tabei was able to prevent these trials lessening her determination.

HONESTY

Climbing Annapurna III, Tabei was frustrated by the inability of her team to be honest about their condition and abilities, even among themselves. "When we began the climb we were determined to only show each other our strong sides," Tabei told a journalist. "When you are climbing a mountain, your life depends on the exact opposite. You can't be reserved and not say what you think or feel." Tabei understood that determination is not to be confused with stoicism; it is about doing what needs to be done, even if that means sometimes admitting difficult truths.

SELF-SUFFICIENCY

GUÐRIÐUR THE FAR TRAVELLER

WATCHING ATLANTIC WAVES POUND THE SHORES AROUND THE ISOLATED VILLAGE OF ARNARSTAPI ON THE SNÆFELLSNES PENINSULA OF ICELAND, IT IS EASY TO UNDERSTAND THAT SURVIVAL IN SUCH AN UNFORGIVING ENVIRONMENT AT THE EDGE OF CIVILIZATION IS DEPENDENT ON UNFAILING SELF-SUFFICIENCY. A THOUSAND YEARS AGO, A YOUNG GIRL FROM THE NEARBY FARM OF LAUGARBREKKA SAILED FROM THIS BRUTAL COAST AND OFF THE EDGE OF THE KNOWN WORLD – TO DISCOVER A NEW ONE.

We only know of Guðríður Þorbjarnardóttir because of the Icelandic sagas, a collection of stories that were written down in the thirteenth and fourteenth centuries but which record much older oral histories. They focus on events that occurred 200 years earlier, during the height of the Viking Age, when Iceland was first settled by the Norse. Originally dismissed as fictional folklore, the sagas are sprawling tales of rich drama and fantastical characters. Guðríður features prominently in two of these books: the *Saga of Erik the Red* and the *Saga of the Greenlanders.* Together, they are known as the Vinland Sagas because they tell of the discovery of new land, far to the west, that the Norse named Vinland, but which we would recognize today as North America. It was only as recently as 1961, when archaeologists uncovered the remains of a Norse settlement in a remote corner of Newfoundland (proving that North America had indeed been discovered by Vikings some 500 years before Columbus) that the sagas were recognized as being, for all their hyperbole and talk of ghosts, valuable texts containing far more historical truth than previously thought. The Vinland Sagas have preserved considerable detail about the first Norse expeditions to the

New World, even if they present very different and often contradictory versions of events and are frequently difficult to interpret. One thing they indisputably agree on is that conspicuous among these extraordinary early explorers to North America was Guðríður.

The sagas tell us precisely where Guðríður was raised in the last decades of the tenth century; on the settlement farmstead of Laugarbrekka belonging to her father, which was – and still is – forbiddingly isolated. Life would have been a harsh existence, challenged by bitter winters and destructive storms, relentlessly stalked by the ever-present threat of famine and illness. With no one to rely on but herself, Guðríður, like all settlers, would have developed from an early age the self-reliant mentality needed in order to survive. Whereas most of us today might find the constant anxiety of being so vulnerable and alone to be intolerable, to the Norse it was an expected way of life. Unafraid of isolation and hardship, they were willing to take the risk of survival in new lands if it offered the opportunity of a better future.

Around the turn of the century, when Guðríður must have been still a young teenager, she sailed from Arnarstapi with her husband and her father, leaving behind everything they knew to join a small community colonizing land that had recently been discovered and settled by a friend of her father's, Erik the Red. Situated to the west of Iceland, Erik called this new country Greenland. The Vikings were expert seafarers, but the voyage from Iceland across the Greenland Sea remained a dangerous one. With no maps or compass, navigation was largely a matter of luck and weather. A decade earlier, when 25 ships had set out with Erik the Red to found the new Greenland colony, only half had landed safely. Guðríður was to suffer a similar fate. Tossed by storms, the ship was disastrously wrecked on a skerry in the middle of the ocean in sight of nothing but an ominous cap of ice. There was no hope of any passing ships in so remote a place – and yet, miraculously, they were saved. Leif the Lucky, son of Erik the Red, had spent the previous winter exploring west of Greenland, investigating rumours of

Top: The wild coast of the Snæfellsnes Peninsula in Iceland, near to the farm where Guðriður was born.

Above: The replica church built close to Qassiarsuk in southwest Greenland in what is thought to be the location of Brattahlið.

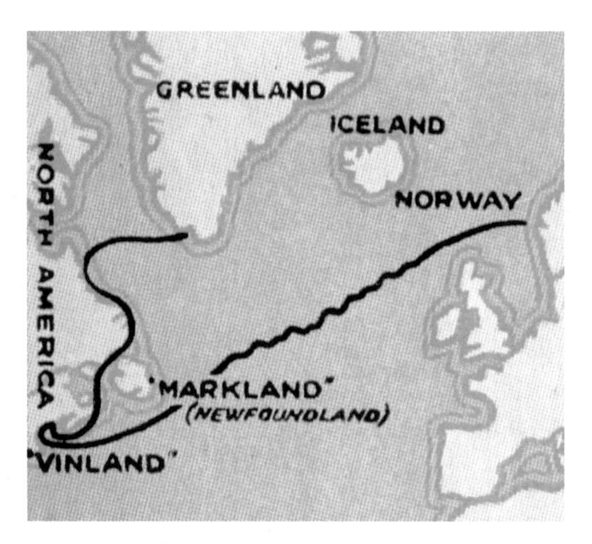

Illustration from 1933 of Guðriður's voyage from Brattahlið in Greenland to the New World and return to Norway.

yet more new territory. He was returning triumphantly to Greenland with news of Vinland when he happened across the shipwreck. Fifteen people were rescued from the skerry, though Guðríður would be the only survivor.

Life for Guðríður at the settlement of Brattahlið on Greenland was even more precarious than it had been in Iceland. Despite good arable lands and pastures for livestock, the short summers were a race against time to prepare as much as possible for the long Arctic winters, when food was as scarce as the daylight and the community would be perpetually on the brink of starvation. What was worse, the normally indefatigable confidence that the settlers derived from their resourcefulness was being undermined by a crisis of faith. The Viking world was undergoing a traumatic shift away from the Norse gods and towards the new religion of Christianity. When hardship struck, many saw it as punishment for betraying the old beliefs, and resorted to traditional ways of magic to protect themselves. Guðríður arrived in Greenland as a Christian but, recognizing the need within the community for hope and reassurance, agreed to revive her knowledge of pagan rites and magic spells in a desperate attempt to ward off calamity. The episode is an insight into Guðríður's strength of character. Described as wise, where wisdom for the Norse was defined by the ability to thrive, come what may, Guðríður was respected for her capability and aptitude for resilience. Expert management of both time and resources during her years at Brattahlið lent Guðríður the practical self-sufficiency so admired by her contemporaries, but also greater confidence to think of a future, to have ambition and aspirations. It gave her an independence of thought that was remarkable enough at the time to be recorded in the sagas 200 years later.

This notable independence perhaps explains Guðríður's decision to make an audacious voyage, sacrificing the safety provided by the colony in Greenland to repeat Leif the Lucky's journey to newly discovered Vinland, where unimagined hardships might wait for them.

She sailed with Leif's brother, Þórsteinn, as his wife, but the couple's ship was blown off course by storms and they were forced back to Greenland to spend the winter in a remote farmstead to the north of Brattahlið. The farmstead was struck by plague during the confined winter and, once again, Guðríður was one of the only survivors. She returned to Brattahlið a widow and, most likely, a woman of some means. Women in the Viking Age did not enjoy anything resembling equality, but they did have some rights that gave them significantly more options than in many other European societies at the time. They were allowed to divorce, to own property, to have custody of their children and reclaim their dowry. Women could also inherit the property and authority of their husbands when widowed. Guðríður, now twice widowed and having already lost her father, would have inherited their wealth, including their ships. Perhaps this is why, when she does finally set sail for Vinland with her third husband, a high-born and heroic Icelandic trader known as Karlsefni, she is presented in the sagas more as a business partner and co-leader of the enterprise than as a wife. This was a large expedition

The excavation of the Oseberg ship discovered within a Norwegian burial mound in 1904. The oak vessel dates from the ninth century and is the oldest as well as best-preserved Viking ship ever found.

of 60 men and five women, and we are told that it was Guðríður who convinced Karlsefni to take part in the risky venture.

Despite many clues in the detailed descriptions preserved in the sagas of the voyage west, we do not know exactly where Vinland would be in modern-day North America. What we do know is that Guðríður and the Norse expedition did not stay in one place. They moved around, exploring the land they had discovered, looking for what it might provide them and learning new ways of survival in an alien landscape. Evidence has been presented that factions of the group may have ventured across modern-day Newfoundland, New Brunswick and the Saint Lawrence River, and perhaps as far south as Rhode Island, Cape Cod or even Manhattan.

As the would-be colonists prepared for the test of their first winter in the New World, Guðríður gave birth to a child, named Snorri, her son with Karlsefni. He was the first European to be born on North American soil and it is testament to Guðríður's resourcefulness that both she and the baby survived and prospered in such a tenuous situation. Guðríður displays the same independence of mind and spirit that had served her so well all her life as she navigates her new reality, including first contact between the European Norse and the indigenous cultures they encountered. The settlers called the indigenous peoples *skræling* and while interactions were initially cordial, even leading to profitable trading, the relationship soon soured. After three years, Guðríður and Karlsefni decided to retreat back to Greenland, most likely due to increasingly violent confrontations with the *skræling*.

Having already made some of the most unprecedented voyages in history and experienced how fatal venturing onto the sea could be, Guðríður had such confidence, in both her husband's ability as a sailor and her own ability to endure, that the couple sailed directly from Greenland to Norway – an ocean journey as long and as arduous as that to the New World. And this time, she had a young child in tow – their young son, Snorri. The sagas describe that the family arrived in Norway with one of the most valuable cargos ever seen, the fruits of the Vinland expedition, which made them simultaneously enormously wealthy and highly respected. In returning from the New World, they had not just survived, they had triumphed. Sailing on to Iceland, the family bought lands in Karlsefni's native region of Skagafjörður in the far north. Guðríður gave birth to another son, and when Karlsefni died some time later, it is perhaps no surprise that Guðríður became head of the family, taking up the authority of her husband and running what would become, under her leadership, one of the most powerful and well-respected estates in the country.

Yet we have to conclude that the life of a high-ranking landowner didn't suit Guðríður. As soon as Snorri was old enough to take responsibility for the family estate, she took the opportunity to reclaim the simpler, self-sufficient independence of her past. Sailing once more across the Atlantic, Guðríður headed south to make a pilgrimage on foot, walking the breadth of Europe to reach Rome, where she made confession, as demanded by her Christian faith, and received absolution from the Pope. Such a journey of religious devotion was encouraged in the eleventh century, but it required resourcefulness and nerve, and it was particularly rare that a woman would make the journey alone as Guðríður had done. By the time of her return to Iceland, she is deservedly believed to be the most travelled woman of the Middle Ages: The Far Traveller.

In the highest gesture of esteem bestowed by the Viking Age, it is Guðríður's family lineage that is listed at the end of *The Saga of Erik the Red*. Guðríður's many descendants are distinguished in the history of Iceland, and her self-contained confidence in her own capability is a characteristic that can still be discerned in the national psyche of modern-day Iceland. Guðríður took holy orders and, as a Christian nun, lived out the rest of her life as a hermit on the family estate, her simple existence exemplifying the self-sufficiency of both thought and deed that had brought her adventure, wealth and, ultimately, the most precious accomplishment of all: fulfilment.

BE AS SELF-SUFFICIENT AS GUÐRIÐUR

SELF-SUFFICIENCY MAY NOT BE A MATTER OF SURVIVAL FOR MOST OF US AS IT WAS FOR GUÐRIÐUR, BUT IT IS STILL A MENTALITY THAT CAN LEAD TO GREATER CONFIDENCE. SHEDDING NEGATIVE DEPENDENCIES ON OTHER PEOPLE AND ON MATERIAL THINGS CAN PROVIDE THE FREEDOM TO FOCUS ON WHAT MATTERS AND TO STRIVE HARDER FOR WHAT WE REALLY WANT.

The inspiring statue of Guðriður carrying her son Snorri, created by Icelandic sculptor Ásmundur Sveinsson in 1938. Copies now stand in Iceland at her birthplace and at the farm where she lived out her days as a nun.

BECOME AN EXPERT (ON YOUR OWN LIFE)

Guðríður was considered wise because of her thorough practical knowledge. Make a point of knowing the practical mechanics of your own life, from big issues such as finances to the more mundane matters of everyday technology, enabling you to build confidence and be free to focus on greater ambitions.

TIME MANAGEMENT

Good management of resources is key to self-sufficiency. Today, as a thousand years ago, time is our most precious resource and often our most stretched. By taking a considered look at how we use our time, we can often find ways to prioritize more wisely.

VALUE SIMPLE THINGS

It is important to focus on what you have rather than on what is missing. It is easy to become reliant on modern technology and material wealth, but the wisdom at the heart of Guðríður's story is that simplicity can be the key to developing practical and personal independence.

HAVE FAITH IN INSTINCT

When deciding to revive her pagan skills at a time of crisis in Greenland, Guðríður prioritized the needs of the colony over her own strongly held beliefs. Then and throughout her life, she made choices that eschewed convention and were instead guided by her own judgement of what was right. She trusted her instinct and it gave her the independence of thought that enabled her to achieve so much.

PATIENCE

CHARLES DARWIN

IN HIS AUTOBIOGRAPHY OF 1882, CHARLES DARWIN STATED THAT HE WAS NOT A CLEVER MAN. DEEMING HIS NATURAL INTELLECTUAL ABILITIES TO BE "MODERATE", HE RATED THE PERSONAL ATTRIBUTES THAT HAD, AS HE SAW IT, ENABLED HIM TO SURPASS THE ORDINARINESS OF HIS MIND AND ESTABLISH ONE OF THE FOREMOST SCIENTIFIC PRINCIPLES OF ALL TIME. SECOND ONLY TO HIS LOVE OF SCIENCE, DARWIN BELIEVED HIS MOST IMPORTANT QUALITY TO BE "UNBOUNDED" PATIENCE.

Darwin's capacity for patience was needed almost immediately on accepting the position of gentleman naturalist aboard the now legendary survey ship HMS *Beagle*. He arrived in Portsmouth having been told to expect the ship to depart on its voyage around the world by the end of September 1831. Instead, he was to learn one of the most inevitable realities of exploring: that even the most carefully laid plans change unpredictably. The departure of the *Beagle* was delayed for a variety of reasons (including the drunkenness of the crew over Christmas) and the ship finally set sail three months late, on 27 December.

The official aim of the voyage was to survey the east coast of South America, gathering information needed to compile marine navigation charts. However, Captain Fitzroy, the man in command of the Royal Navy expedition, recognized an opportunity for a wider exploration of the remote territories that the ship would encounter and engaged Darwin for the purpose. Initially, the expedition was expected to take two years, but it would be 1,740 days, almost five years, before the *Beagle* returned to England. After sailing along the coast of South America and navigating Cape Horn, the ship transited the Pacific Ocean to Australia, New Zealand and Cape Town before heading home in 1836.

The robust but cramped *Beagle*, which was the crew's home and workspace for all this time, was just 30 m (90 ft) long. Darwin shared a confined cabin with three others, sleeping at night in a hammock suspended over the table where he worked – a cabin that already doubled as the ship's crowded library, containing some 400 volumes. Although much of his time was spent at sea and aboard the ship, Darwin took every opportunity to travel inland wherever the ship ventured. Setting off on daring journeys, often accompanied only by the sailor he had trained as his servant and assistant, he enthusiastically investigated geology, fauna and flora that had never before been recorded. Everywhere he went he collected samples, thousands of them, selecting the most interesting to be crated home whenever the ship called at a major port. The samples were received by an array of specialists in various disciplines, whose expert help Darwin had recruited in order to identify and assess his specimens.

Of birds alone, Darwin gathered 500 skins as well as nests, eggs and whole birds preserved in spirit. Yet, despite the volume of his zoological collection, Darwin's first priority on the *Beagle* was geology. He collected three times as many geological samples as he did anything else, including masses of dinosaur bones, many of which were identified back in England as species new to science. The finds made Darwin famous, even in his absence, so that by the time he returned from the expedition he was already a scientist-collector of considerable reputation. Nevertheless, these collections were not to be Darwin's greatest achievement. As history now records, his explorations planted in his mind the seeds of an idea that would grow into one of the most important theories of science: evolution by natural selection. But it would take another 20 years of patient study and diligent, deep thinking before Darwin considered the concept that was destined to change the world to be fully formed.

Watercolour by expedition artist, Conrad Martens, showing HMS *Beagle* at Tierra del Fuego, the region at the southernmost tip of South America.

Sketch of *Mylodon Darwinii*, 1890. This species of extinct ground sloth was discovered from bones collected by Darwin at Punta Alta, Argentina in 1832 and later named for him.

Darwin had joked in correspondence with a friend on his return to England that the materials from the *Beagle* expedition would "take twice the number of years to describe, that it took to collect & observe". And so it proved to be. By the end of 1846, a decade after his homecoming, Darwin was still fully occupied in a prolific publishing and editing campaign of his recollections and scientific work from the expedition. Between 1839 and 1846, he produced 10 major books – including several colossal volumes of zoology as well as publications discussing substantial topics such as coral-reef formation, volcanic islands and South American geology – in addition to more than 20 acclaimed contributions to periodicals. It is a staggering output, and it is hard to believe that he could have had time to work on anything else at the same time, never mind a theory as complicated and intricate as natural selection. In the extensive notebooks that Darwin used to meticulously record all his observations during the voyage of the *Beagle* – a habit he maintained for the rest of his life – the framework of his ideas on evolution and the diversity of nature were already mapped out in some detail by as early as 1839. Despite sensing that his ideas were important, Darwin resisted publishing, deciding to wait so that he could think about all aspects of his argument, probing any weaknesses he perceived, wanting to be satisfied that his hypothesis was as rigorous as possible. "I suppose that I am a very slow thinker, for you would be surprised at the number of years it took me to see clearly what some of the problems were, which had to be solved," he wrote later in his life. "Looking back, I think it was more difficult to see what the problems were than to solve them..."

Darwin set about industriously unearthing supporting evidence for his theory whenever he could, but he considered these to be his personal investigations, and the punishing workload of the *Beagle* publications forced them into the background. When he finally started working on invertebrates, the last zoological category from the *Beagle* voyage to be studied, Darwin could have been forgiven for hastening his progress in order to have more time to devote to what he termed his "species" work, but he refused to sacrifice any of his conscientiousness. The planned volume on invertebrates gradually extended into an exhaustive eight-year study of barnacles, another stunning exercise in thoroughness, which was finally published in 1854.

Darwin did occasionally express doubts about the value of his investigations of barnacles. Ultimately, though, he viewed this time-consuming study, and the 10 years of diligent work on his material from the *Beagle* that preceded it, not as a distraction that delayed his important work on species, but as time that had been essential to the development of his evolutionary theories. The *Beagle* investigations that had kept him so occupied, had each subtly changed his thinking in important ways. His research on barnacles gave him a thorough knowledge of taxonomy (the science of naming, describing and classifying) and of morphology (the study of the form and structure of organisms), both of which added greatly to his understanding of natural variation. Darwin had previously assumed that species originated in distinct geographical centres, but he began to comprehend through his examination of barnacle species a more universal process of selection.

Similarly, Darwin's study of geology, which might appear on the surface to have had little in common with zoology, lent him the concept of "gradualism" – the understanding that slow and incremental processes could accumulate over vast periods of time to produce significant changes. It is a notion so intuitive to us today that, much like trying to imagine a time before our knowledge of inheritance through genes and DNA, it can be hard to appreciate the depth of thought and the time necessary to establish such revolutionary thinking. "This multiplication of little means and bringing the mind to grapple with great effect produced is a most laborious & painful effort of the mind," Darwin wrote, hinting at the patience and precision required to determine all the small but consequential components leading to the diversity and variation of species he had observed. Darwin came to recognize that the scientific explanation of species would reveal itself only through a mass of observations and the painstaking study of abundant specimens of as many different organisms as possible. It was an overwhelming task, but one that Darwin tackled methodically and relentlessly for decades. Not only did he gather specimens of as many species as possible, investigating their development and variation, but he also began experimenting with breeding animals and plants, reading widely and consulting with experts in animal husbandry, from farmers to pigeon fanciers. He studied the dispersal

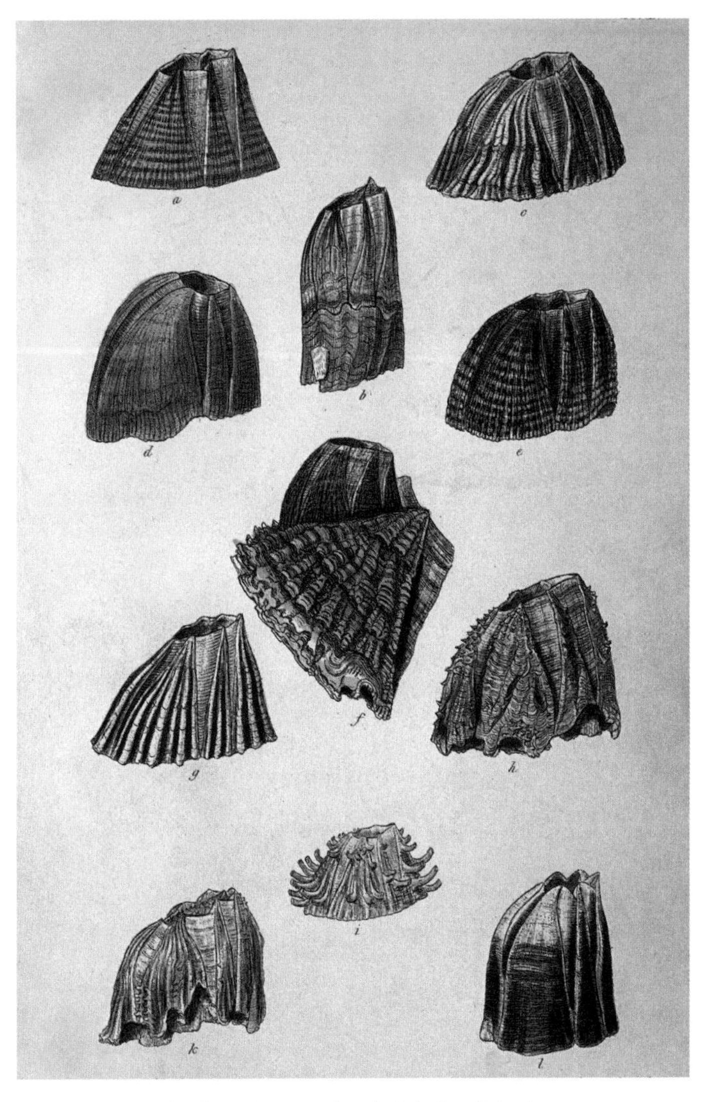

Plate from *A Monograph on the Sub-Class Cirripedia, With Figure of All the Species* published by Darwin in 1854 after an exhaustive eight-year study.

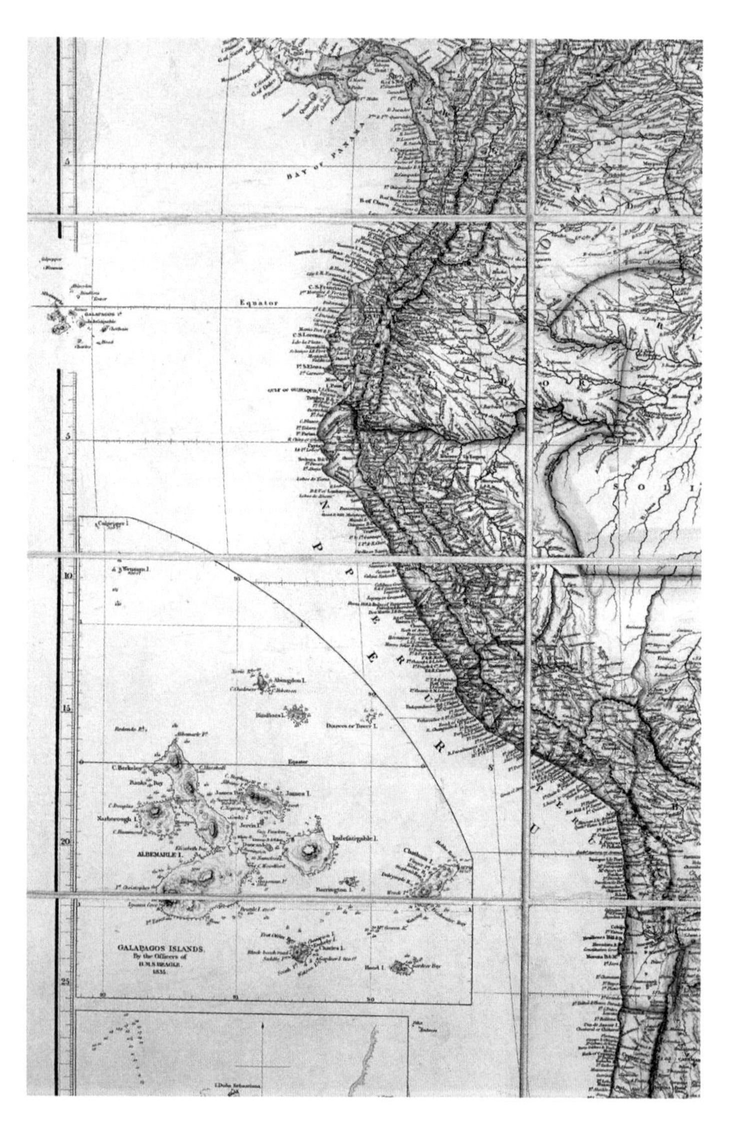

Pen and ink chart of the Galapagos, an archipelago located off the east coast of South America, compiled by officers of HMS *Beagle* in 1835.

of seeds, spent time probing the formation and development of the geometric structure of beehives, and cultivated a network of colleagues and friends who would not only funnel new information to him, but also act as a critical audience on which to test his ideas.

As well as the sheer scale of the task he had set himself, there were other draws on his patience. Before travelling around the world, Darwin had suffered bouts of debilitating illness that had no clear diagnosis. On his return to England, while still a young man of 27, his condition gradually became chronic while its cause remained a mystery. Stomach complaints, headaches, palpitations, eczema, boils, retching and exhaustion plagued him almost continually alongside terrifying psychological symptoms that he described as, "shivering, hysterical crying, dying sensations or half-faint ... singing of ears, rocking, treading on air, focus & black dots". The baffling affliction left Darwin in agonies of both mind and body, unable to work for more than two or three hours a day and sometimes not at all for months at a time, a situation that drove him to even greater anxieties. The symptoms remained as persistent, without either cure or explanation, for the rest of his life. "These causes combined have given me the patience to reflect or ponder for any number of years over any unexplained problem," he wrote in his memoirs. He comforted himself with the thought that his ill health had kept him away from social distractions, leaving more time to devote to study, and was grateful that even when his affliction had stopped him working, he had still been able to think on his "species" manuscript.

On the Origin of Species by means of Natural Selection, or the Preservation of Favoured Races in the Struggle for Life was eventually published in November 1859, followed by a further six editions in Darwin's lifetime. Such a length of time between first thought and published theory was far from unusual; it was a feature of much of his work. The study of the psychological development of his baby son, William, was not published for 37 years. Similarly, there were 37 years between the start of his work on cross-fertilization and the publication of his book on the subject. His observations on orchids were not published for some three decades and in 1837, when he began investigating how the seemingly insignificant action of earthworms were able to greatly alter the landscape, he probably did not anticipate that it would be 42 years before he published his findings. Having the patience to let his ideas mature and develop was a key feature of Darwin's prolific career.

Darwin did not discover evolution. The idea had been discussed in various forms by a number of academics before him, including his own grandfather. Darwin's unique achievement was in conceiving both natural and sexual selection, as well as in collating evidence from far-reaching sources to produce an unassailable argument. Rather than rushing to publish his concept of natural selection in 1839, he crucially had the patience to wait, giving himself the time to analyze every aspect of his theory and develop his understanding until he had a complete explanation for every shred of evidence, every apparent anomaly. He was described by his granddaughter as a "patient collector of facts", and it seems the gradualism that infused Darwin's scientific ideas also inspired his method of thought. The result was a discovery that is a fundamental tenet of our understanding of life and humanity and which now bears his name – Darwinism – a discovery that changed not only *what* we thought, but *how* we thought, forever.

Scribbled sketch from Darwin's Notebook B, drawn in 1837, that would form the basis for his later theories on evolution. The most ancient life forms are shown at the base of the tree, with descendants appearing as branches.

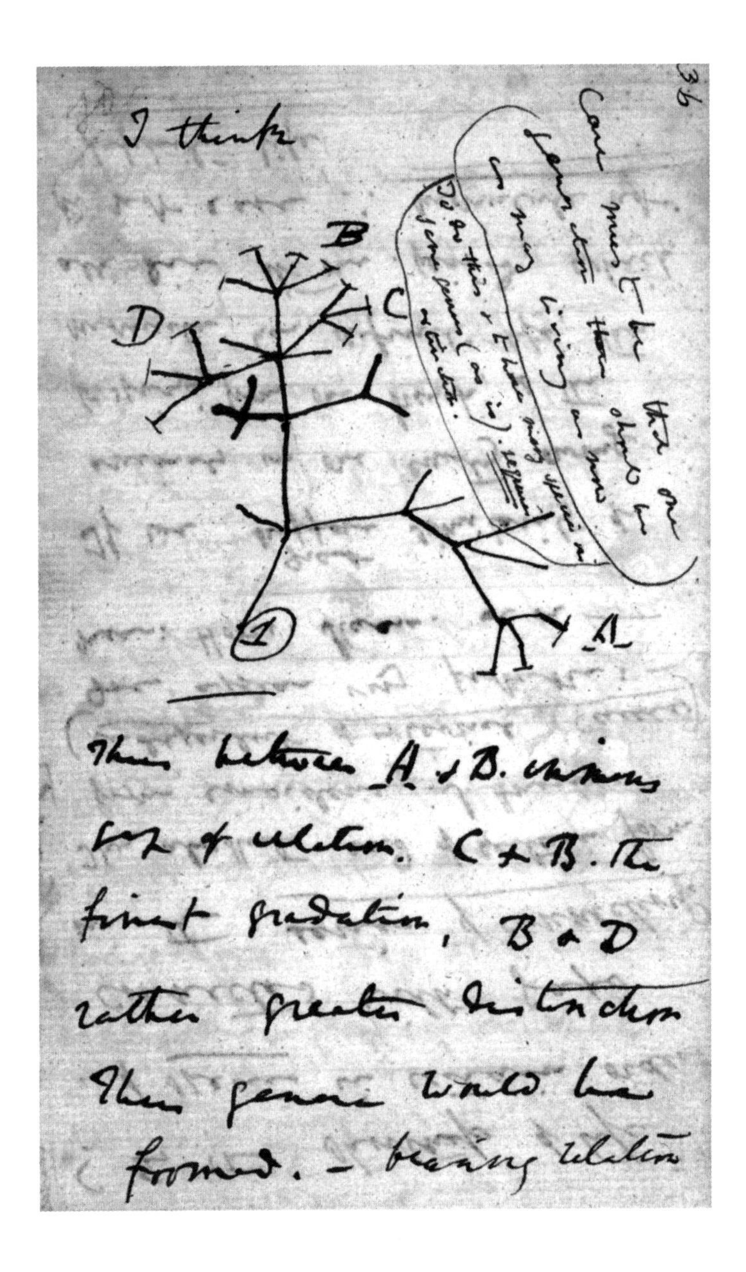

36

I think

Case must be that one generation then should be as many living as now. To do this & to have as many species in same genus (as is) requires extinction.

Thus between A & B. immense gap of relation. C & B. the finest gradation, B & D rather greater distinction. Thus genera would be formed. — bearing relation

DEVELOP DARWINIAN PATIENCE

TO HAVE PATIENCE REQUIRES DISCIPLINE OF BOTH THOUGHT AND ACTION. WE ARE FORTUNATE THAT DARWIN HIMSELF SPENT TIME DETERMINING THE ATTRIBUTES AND HABITS THAT ENABLED HIM TO BE SO SUCCESSFUL AT MAINTAINING HIS LIFELONG DILIGENCE AND THAT HE THOUGHT TO SHARE THESE INSIGHTS IN HIS WRITINGS SO THAT OTHERS MIGHT MAKE USE OF HIS STRATEGIES.

Ceroglossus beetle collected by Darwin from the Chilean Andes during the HMS *Beagle* expedition. One of thousands of specimens Darwin gathered which remain of great scientific value today.

PRIORITIZE WEAKNESS

What Darwin termed "prejudice" we would today call "confirmation bias" – the inclination to see only information that supports the preferred action or thought. Darwin strived to eliminate prejudice by deliberately seeking out opposing evidence, welcoming the opportunity to analyze any problems directly and immediately: "I had, during many years, followed a golden rule, namely, that whenever a published fact, a new observation or thought came across me, which was opposed to my general results, to make a memorandum of it without fail and at once; for I had found by experience that such facts and thoughts were far more apt to escape from memory than favourable ones. Owing to this habit, very few objections were raised against my views which I had not at least noticed and attempted to answer."

EXTREME OBJECTIVITY

Darwin warned against becoming emotionally attached to favoured ideas, understanding how easy it is to allow ego invested in a project to blind us to faults and weaknesses. "I have steadily endeavoured to keep my mind free so as to give up any hypothesis, however much beloved (and I cannot resist forming one on every subject), as soon as facts are shown to be opposed to it," he said.

TAKE NOTES

The notebooks that Darwin filled during the voyage of the *Beagle* and the years that followed record not just field observations but impressions from his travels and life, as well as jottings from texts and discussions. Such a meticulous chronicle of his thoughts and ideas meant that Darwin was better able to distinguish gaps in his understanding and could capitalize on moments of insight and inspiration he had recorded, many years after they had first occurred.

CURIOSITY

ABU AL-HASAN AL-MASUDI

AT THE TURN OF THE NINTH CENTURY, BAGHDAD WAS A CITY THAT SPECIALIZED IN KNOWLEDGE. IT WAS THE CAPITAL OF THE MUSLIM WORLD AT THE HEIGHT OF THE GOLDEN AGE OF ISLAM, AN ELECTRIFYING CONCENTRATION OF THE GREATEST SCHOLARS AND THINKERS, A PRODIGIOUS CENTRE OF INTELLECTUAL LEARNING AND HOME TO A VAST LIBRARY OF MANUSCRIPTS KNOWN AS BAYT-AL-HIKMA: THE HOUSE OF WISDOM.

However, for one ambitious young academic who had been born and raised in the city, this ready access to an abundance of knowledge was not enough. The erudite environment of Baghdad trained Abu al-Hasan al-Masudi in the art of learning, but only served to excite in him a greater thirst for information and a curiosity that he determined could never be satisfied by a lifetime of scholarly inquiry. Instead, he demanded personal, first-hand experience. Rather than learning about the world from books, he resolved to go out into the world and put what he learned into a book of his own.

To travel, not for trade or as a diplomat or to make war, but solely to learn, was unheard of at the time. For all its love of knowledge, Baghdad was a fundamentally insular society. Learning of people and places outside of the Muslim world, in both time and geography, was not particularly encouraged or thought to be of any great intellectual worth. As in most European societies of the age, severe religious prejudice and mistrust of the unfamiliar was deep-rooted. Non-Arabs and non-Arab Muslims experienced a tenuous acceptance in Baghdad society, but their presence was frequently volatile. Masudi was doubly unusual: his curiosity was not limited by religion, politics or culture and, in a direct challenge to the traditionalism of the day, he considered experience to be of more importance than the authority of intellectual writers. Leaving the rich academic resources of Baghdad behind, Masudi embarked on a lifetime of exploration and discovery, describing himself as "avid to learn for myself all the remarkable things which exist among different peoples and to study the particular characteristics of each country with my own eyes".

Guided by his wide-ranging curiosity, and making use of the extensive Arabic trading routes that extended outwards, tentacle-like, from Baghdad, he first ventured eastward. Travelling through Persia and modern-day Pakistan, he went beyond the domains of the Muslim world into India, traversing the high central plateau of the subcontinent and, on reaching its southern tip, sailing to Sri Lanka. Probing ever further east, he explored what is now Indonesia and Malaysia before journeying to the southern reaches of China – the very edges of the medieval trading world, where few had personally ventured. Returning across the Indian Ocean, he visited Madagascar, Zanzibar and the east coast of Africa, including the regions known today as the countries of Mozambique, Tanzania and Kenya. On returning to the Arabian Peninsula via Oman, he was led west, crossing into Palestine and eventually following a caravan route across the Syrian Desert to the ancient city of Palmyra and into what would become western Iraq. From there, he turned his attention to the north, roaming throughout the Caucasus and the regions surrounding the Caspian Sea into Central Asia before coming back to Arabia. After taking some time to complete a pilgrimage to the holy Islamic city of Mecca (today in Saudi Arabia), he continued to Yemen and then to Egypt. Masudi was now an old man of around 50 years of age, and would spend the last decade of his life moving back and forth between Egypt and Syria. Travelling almost until his last breath,

Thirteenth-century illustration of scholars at the House of Wisdom in Baghdad, a repository of manuscripts gathered by Muslim rulers during the Golden Age of Islam.

he ended his days in a city just outside modern-day Cairo. For the entirety of his adult life, Masudi had toured most of the world known in the Middle Ages, so it was no exaggeration when he claimed: "I journeyed from one quarter of the Earth to the other as the sun makes his eternal revolutions."

What Masudi observed on his travels, he put in his books. Although the majority of the 38 volumes he wrote during his life have been lost – volumes with titles as simultaneously ambitious and intriguing as *The History of Time* and *The Secret of Life* – the rest survives as a single epic work known as *Meadows of Gold and Mines of Gems*, with an intimidating 132 chapters. The latter half of the book gives a detailed history of Islam, from the prophet Mohammed and the dynasty of his descendants who became successive caliphs of Islam, right up until 957 CE, the year of Masudi's death. The first half contains everything else about the world and human existence that Masudi had observed, and is where the full elasticity of his curiosity is most gloriously revealed. In the introduction, he discloses that the book encompasses, among many other things,

> the figure of the Earth, its towns, wonders, and seas, its heights and depths, mountains and rivers, the produce of the mines; the various waters, marshes, and the islands of the sea and of lakes. I have also given descriptions and historical sketches of large edifices and lofty temples, an account of the beginning and last origin of things, and notices of inhabited districts, and of such tracts as had been land and became sea, or which had been sea and became land; together with the causes of those changes, both proceeding from cosmic and natural influences.

In surely the most eclectic chapter index ever committed to paper, Masudi gives free rein to the vastness of his experience. In a great amalgamation of geography, history and the natural, human and physical sciences, he jumps from the creation of man to the

Israelite kings and prophets, Hindus, lunar tides, the Mediterranean and Caspian Seas, the solar system, climatic regions, mountains, great rivers, the Chinese Empire, the Caucasus, the first Persian kings, the Greeks, the history of Alexander the Great in India, the Roman Empire, Christian sovereigns of Byzantium, Egypt, Sudan, the Slavs, Franks and Galicians, Germanic peoples, Yemen, the kings of Syria, Kurds, nomadic Arabs, pre-Islamic beliefs, the Arabic diaspora, a history of the elephant, ghosts and witchcraft, soothsaying, a comparison of various calendar systems, sacred temples, and perhaps most intriguing of all, an entire chapter dedicated to "ominous sounds".

Not only did Masudi draw from his own encyclopedic observations, but he also took great pains to personally seek and consult other knowledgeable scholars in the lands he visited. At a time when foreign cultures were largely considered inferior and thought to have little understanding of the world that was of any value, Masudi instead uniquely perceived an opportunity to vicariously expand his knowledge. For example, by examining the work of those who would be termed "local collaborators" in the parlance of modern-day exploration, Masudi could accurately relate of the Arctic, a region he never personally visited, that "the inhabited area extends from the equator as far north as Thule, beyond Britain, and there the longest day lasts 20 hours."

This open-minded attitude towards unfamiliar and unknown peoples and lands is a hallmark of his work, which reveals as much about the character of Masudi as it does of the world he explored. Nothing appears to have been beneath his attention. He finds as much of interest in the stories of slaves and prisoners, merchants and market traders as he does in the lives of kings, and is as likely to recount the intricacies of diving for pearls in the Persian Gulf or the recipe of a favourite dish from the caliph's table as he is to describe the course of

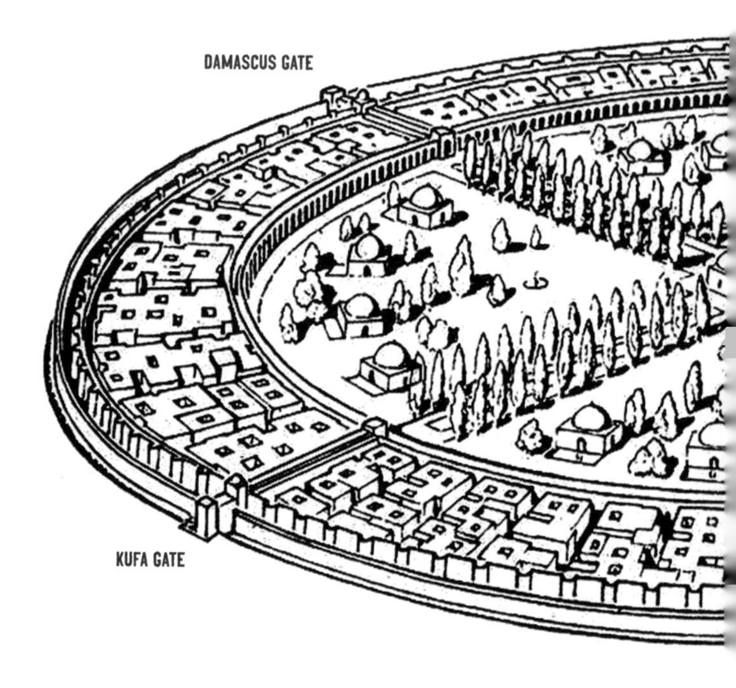

the Nile or the land route to China. "I have called attention to those subjects which the vigilant observe, and upon which the intelligent reflect, and to what they mention of the splendour which enlightens the world," he wrote.

His far-reaching curiosity extended to gender, with women of all ranks in society not neglected in his narratives and often taking centre stage. His curiosity about other religions also seems to have been strikingly free of prejudice in comparison to the conservative attitudes of the day. In Persia, he makes a point of visiting and understanding the temples and customs of Zoroastrianism; in India,

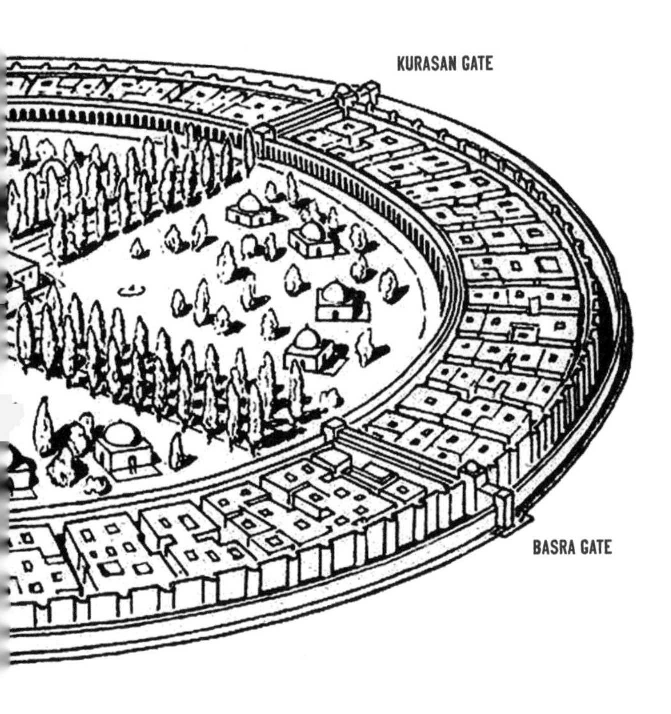

The Round City of Baghdad, completed in the mid-eighth century, would have formed the core of the city known by Masudi, a venerated centre of culture and learning.

he is fascinated by Hindu burial customs and records them in detail; in Palestine, he visits the sites of importance to Christianity and spends time in discussion with Christian nuns. Judaism, polytheism, heretical Muslim sects, pre-Islamic Arabic beliefs and the religions of China are all similarly treated with scholarly attentiveness. Masudi declares, "May the reader rest assured that I have not here taken up the defence of any sect, nor have I preferred this doctrine or that. My aim has been to trace the beauty of the histories of many peoples, and I have no other."

This broadness of vision gave his writing a vibrancy that set it apart from other academic texts and allowed him to combine widespread influences and knowledge to perceive new and exciting ideas. He hypothesized on the causes of earthquakes, was the first to record that the Caspian and Black Seas were not connected, and developed a theory of life mutating from minerals to plant matter – a concept that presages ideas of evolution that wouldn't emerge for another 900 years. He also recalls seeing a map for the first time, remembering "climates represented in various colours, without a text ... constructed by a group of contemporary scholars to represent the world with its spheres, stars, land, and seas, the inhabited and uninhabited regions, settlements of peoples". In the Middle Ages, Arabic maps existed only as illustrations to academic texts, and the map of the world Masudi was inspired to create is one of the earliest. Placing south at the top, he drew a version of the world's landmasses accurate enough to be recognizable today. It depicts major rivers including the Nile, the Yangtze and the Ganges, as well as the Caspian and Aral Seas which had not been described before. Masudi perceived that maps are more than a simple marking of territory; they are a means of understanding. This greater comprehension of the world was the driving motivation behind his relentless and all-encompassing quest for knowledge.

By insisting on first-hand information and on the assiduous separation of fact from myth, Masudi created a volume of literature which was considered so reliable that it was used as a reference and for verification of other works in his own lifetime and for centuries after. Today, it remains one of the most complete records we have of the ancient world. By introducing analysis, reflection and social commentary to his observations, Masudi pioneered the discipline of scientific

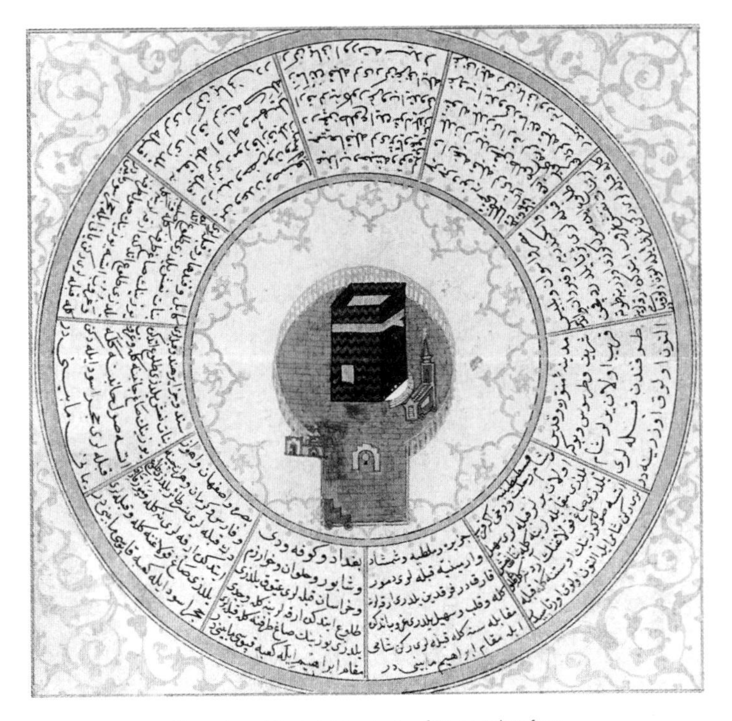

Illustration of the holy Islamic city of Mecca taken from a thirteenth-century Arabic manuscript.

geography and established a grand tradition of field exploration that still exists a thousand years later. In a proclamation that has echoed through the centuries and formed his legacy, Masudi declared, "He who has never left his hearth and has confined his researches to the narrow field of the history of his fatherland cannot be compared to the courageous traveller who has worn out his life in journeys of exploration to distant parts and each day has faced danger in order to persevere in excavating the mines of learning and in snatching precious fragments of the past from oblivion."

CULTIVATE THE CURIOSITY OF MASUDI

TO BE HUMAN IS TO BE CURIOUS. THE NEED TO EXPLORE OUR ENVIRONMENT AND TO ASK QUESTIONS IS WHAT HAS DRIVEN OUR PROGRESS AS A SPECIES AND HAS POWERED THE RISE OF CIVILIZATIONS. MASUDI PROVIDES A STUNNING EXAMPLE OF HOW MUCH WE CAN ACHIEVE WHEN WE ALLOW OURSELVES TO BE GUIDED BY OUR NATURAL SPIRIT OF CURIOSITY.

The world map compiled by Masudi has not survived but this tenth-century Arabic world map dates from around the time of Masudi's death, created by fellow Islamic scholar Abu Ishaq Ibrahim ibn Muhammad al-Farisi al-Istakhri.

PERSONAL EXPERIENCE

Masudi valued first-hand knowledge above all other forms of education, believing that we learn best by doing. He considered getting out into the field to explore through direct experience, as well as the immediacy of seeking greater exposure to the unknown, to be the most effective way of gaining better understanding of people, places and, most importantly, of ourselves.

CHALLENGE YOURSELF

It is easy to get stuck in a rut of the familiar. We can take from the story of Masudi the rewards to be had by pushing ourselves beyond personal comfort zones to broaden our horizons and of having the courage to deliberately pursue what is new, alien and even perhaps, at first, frightening. He showed that it is only by continually trying new things that we can fully explore the extent of our capacity for learning and understanding.

BE OPEN-MINDED

Masudi was ahead of his time when it came to the value he placed equally on all human stories. His wisdom was that there is something to learn from every experience, no matter how seemingly unimportant or irrelevant it might at first appear. He demonstrated that priceless knowledge can come from unexpected sources and that whatever the situation or circumstance, we should be open to the learning experiences that can be gained.

RESILIENCE

DAVID LIVINGSTONE

AT THE OUTSET OF THE NINETEENTH CENTURY, AFRICA REMAINED A VAST UNKNOWN TO THE WESTERN WORLD. EXPLORATION HAD VENTURED LITTLE FURTHER THAN THE COAST OF THE CONTINENT, AND IN THE EUROPEAN IMAGINATION, THE AFRICAN INTERIOR WAS A DANGEROUS PLACE OF CANNIBALISM AND SAVAGERY, OF MAN-EATING BEASTS AND GRUESOME TROPICAL DISEASE. IT WAS VARIOUSLY REFERRED TO AS THE DARK CONTINENT AND THE WHITE MAN'S GRAVE.

Small wonder, then, that as he prepared to set off inland across uncharted territory from the eastern shores of Africa in 1871, a 30-year-old Henry Morton Stanley was terrified. The horrors that he suspected were waiting for him in the African wilderness led him to contemplate suicide in preference. He had been charged by the most popular newspaper of the day with the task of finding the most famous explorer of the age, Dr David Livingstone, who had not been heard from for more than a year on his expedition to seek the source of the Nile. The public were enthralled by the mystery of his apparent disappearance and were desperate for answers. Stanley was a young, ambitious reporter, eager to make his name. A Welshman born into poverty, he had emigrated to the US at the age of 17, taken a new name, and became a veteran of the American Civil War who had fought for both sides. He was undoubtedly tough, shrewd and accustomed to living on his wits – but even he was humbled by the trials of the African bush. He had readied a party of more than 100 people, confidently vowing of his mission to find Dr Livingstone: "If alive you shall hear what he has to say. If dead I will find him and bring his bones to you."

Eight months later, both of his white companions were dead, as was his thoroughbred horse. Even his favourite donkey had been

eaten by crocodiles during one of many treacherous river crossings. For more than 1,200 km (750 miles) he had battled swamps and mountain ranges, fought dementia brought on by cerebral malaria (which was usually fatal), near continuous dysentery and an almost terminal bout of smallpox. Permanently on the brink of starvation, two thirds of his retinue had died or deserted, and having stumbled into the midst of a tribal war, the expedition was treated with suspicion and hostility by anyone they were unlucky enough to meet. The only glimmer of encouragement amid such prolonged misery was the rumour of a white man staying in a village called Ujiji on the shores of Lake Tanganyika. Stanley arrived there surrounded by a crowd of excited villagers and found the missing Livingstone spending his days sat on a goatskin beneath the eaves of a mud hut, reduced to "beggary".

Photograph from 1895 of the southern end of Lake Tanganyika, where 14 years earlier Stanley had found Livingstone in the lakeside village of Ujiji.

> "As I advanced slowly toward him I noticed he was pale, looked wearied, had a gray beard, wore a bluish cap with a faded gold band round it, had on a red-sleeved waistcoat and a pair of gray tweed trousers. I would have run to him, only I was a coward in the presence of such a mob, – would have embraced him, only, he being an Englishman, I did not know how he would receive me; so I did what cowardice and false pride suggested was the best thing, – walked deliberately to him, took off my hat, and said, 'Dr. Livingstone, I presume?'"

Livingstone was close to death. Since setting out from the east coast five years earlier with 35 men, he had traversed hundreds of miles of swamp, mangrove and jungle, tracing a meandering route around south-central Africa searching for clues as to the origins of the waters of the River Nile – but with little success. By the time Stanley found him, Livingstone was accompanied by just three loyal porters, all that remained of his expedition. The majority of his company had deserted early on with large portions of his supplies and later made off with his precious medical kit. As a result, the famously hardy explorer who had once boasted that his muscles were "hard as a board" was emaciated from dysentery and anaemia, crippled by infected ulcers on his legs and had needed to extract almost all of his own teeth due to disease. Having exhausted his funds, he was unable to procure food, shelter or even the most basic of supplies. For months, he had been using crushed berry juice and scraps of newspaper to write letters he had no possibility of sending and, in a further indication of his disorientation, scribbled diary notes with dates that were inaccurate by weeks. As the only European, as far as he knew, for hundreds of miles in any direction, Livingstone had presumed himself given up for dead by the outside world. The arrival of Stanley must have seemed a miraculous salvation. For Stanley, the man he found in Ujiji must have seemed little more than a ghost of the legendary Dr Livingstone.

Perhaps because he had himself now experienced the trials of travelling through the wilds of Africa, Stanley was deeply affected by the elder explorer: a man 27 years his senior who had been exploring – and surviving – the continent for longer than Stanley had been born. Rather than hurry home with the scoop that he knew would make his career and fortune, Stanley oversaw Livingstone's recuperation, and then agreed to accompany him on an expedition across the length of Lake Tanganyika in a dugout canoe. It was a journey Livingstone would describe as "a pleasure trip" but which Stanley saw as a priceless masterclass in exploration from a man who had trekked 47,000 km (29,000 miles) across Africa with a resilience so astonishing that he was perceived as a phenomenon.

Livingstone had first been persuaded to come as a missionary to Africa by the idea that the spread of Christianity – and the European commerce it was believed would inevitably follow – could rid the continent of the devastating trade in human life caused by slavery. At that time, the slave trade was still a profitable industry perpetrated by an established network of predominantly Arab traders. Arriving in 1841, Livingstone began establishing small mission posts at the edge of the Kalahari Desert in what is now Botswana, where he was promptly almost killed by a lion. He described the lion mauling him "as a terrier dog does a rat" and was left with a badly splintered arm that he had to set himself without anaesthetic or the help of another doctor. The arm remained debilitated for the rest of his life. While recuperating he met and later married the daughter of a fellow missionary, and as they moved between mission posts in the Kalahari region, he began his first exploratory travels northward, some of which he attempted accompanied only by his growing family. He became the first European to cross the challenging sand and scrub of the Kalahari Desert, to discover Lake Ngami and to learn of major rivers that ran into Africa's interior, particularly the Zambesi. Livingstone became convinced that these rivers could act as thoroughfares which would open up previously inaccessible regions of Africa to the evangelism and commerce that he believed would supplant the slave trade.

While accompanying him on his explorations across the Kalahari, Livingstone's young family came perilously close to dying of thirst and, later, starvation. At one point his wife Mary, heavily pregnant with her fourth child, suffered extreme paralysis due to the hardship, and the baby, born shortly after the expedition's return, survived just six weeks. In 1852, as Livingstone prepared to return to the Zambesi and determine whether it was a plausible route through the African wilderness to the coast, he made the decision to send his family to

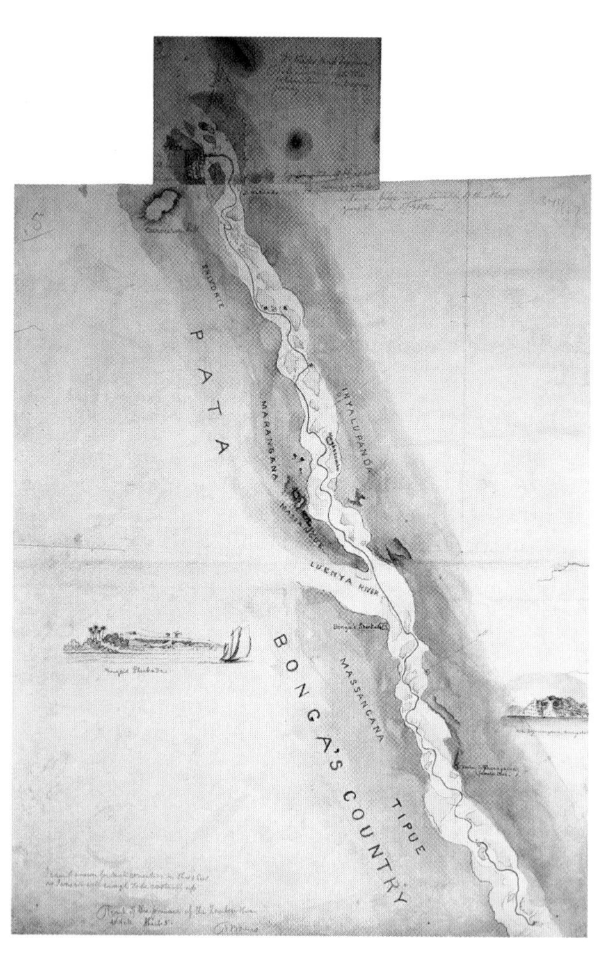

Hand-drawn map from 1859 showing the route followed by Livingstone down the Zambesi on his ill-fated expedition by steamship to prove the navigability of the river.

Britain. By all accounts, he and Mary shared a loving and supportive marriage, and the eldest of their four children was aged just six at the time, so we can assume it was a difficult separation. He would remain a mostly absent figure for the rest of their lives. Yet his subsequent journey along the Zambesi would be Livingstone's great triumph. For the next four years, Livingstone followed the course of the river that snaked for 3,540 km (2,200 miles) through the wilds of the African interior. He tracked first to the Atlantic on the west coast and then to the Indian Ocean in the east, thus becoming the first European to see what he named Victoria Falls and the first European to cross the southern African continent. It is hard to overestimate the sensation caused by Livingstone's account of his discoveries when he returned to Britain. Previously, the outside world had assumed the centre of Africa to be desert, from the Kalahari to the Sahara. The impact can be likened to future explorers of the planet Mars returning to Earth to describe an unexpected world of rich wildlife, exotic civilizations and a verdant landscape. More extraordinary still for his Victorian audience was that Livingstone had survived four years in an environment thought swiftly fatal for white people.

Livingstone credited his upbringing for his seemingly superhuman resilience. Born in Blantyre, an industrial town close to Glasgow, he started working in a cotton mill by the age of 10 and claimed that the long monotonous days taught him persistence and endurance. With just 40 minutes allowed twice a day for meals, Livingstone would work from 6 a.m. to 8 p.m. and then go to school in the evenings. He managed to attend to his education well enough that at the age of 26, he began studying medicine at university. Others have suggested that his success on the Zambesi was due to his attitude towards the people of Africa, which was uncommonly empathetic for the time. He could speak Setswana, a predominant regional language, fluently, and had a developed understanding of tribal customs, beliefs and politics. Still others attributed his survival to his medical knowledge. Indeed, while the understanding of the causes – much less treatment – of malaria, Africa's great killer, was still 40 years away, Livingstone already understood that quinine had the ability to aid recovery from the disease even if it couldn't prevent it. Having contracted malaria at least 30 times during his explorations, he created pills that combined quinine and a selection of purgatives, advising these be taken until they induced a "ringing in the ears". "They received from our men the name of 'rousers,' from their efficacy in rousing up even those most prostrated," wrote Livingstone.

Within 18 months of his return to Britain, Livingstone was on his way back to Africa, accompanied by Mary and their youngest son. The aim of this government-funded expedition was to prove the navigability of the Zambesi by taking a steamship upriver from the coast – but as the impossibility of the mission became clear, Livingstone's resilience took a dangerous turn towards blind determination. The questionable judgement he had exhibited previously in risking the lives of his family in pursuit of his objectives was exacerbated by the repeated failures of the expedition. It led one European team member to record, "Dr L. is out of his mind ... he is a most unsafe leader." His reckless tenacity was worsened yet further in 1862 by the death of his wife.

On finding she was again pregnant, Mary had not travelled along the Zambesi with Livingstone but returned to Britain with her newborn daughter. As the expedition was extended from two to four years, she left her five children behind to travel to Africa to visit Livingstone, only to be claimed by malaria within six months of her arrival. In his grief, Livingstone only increased the frenetic efforts of the expedition, refusing to concede defeat against all reason until he was finally recalled to Britain in 1864 by his increasingly

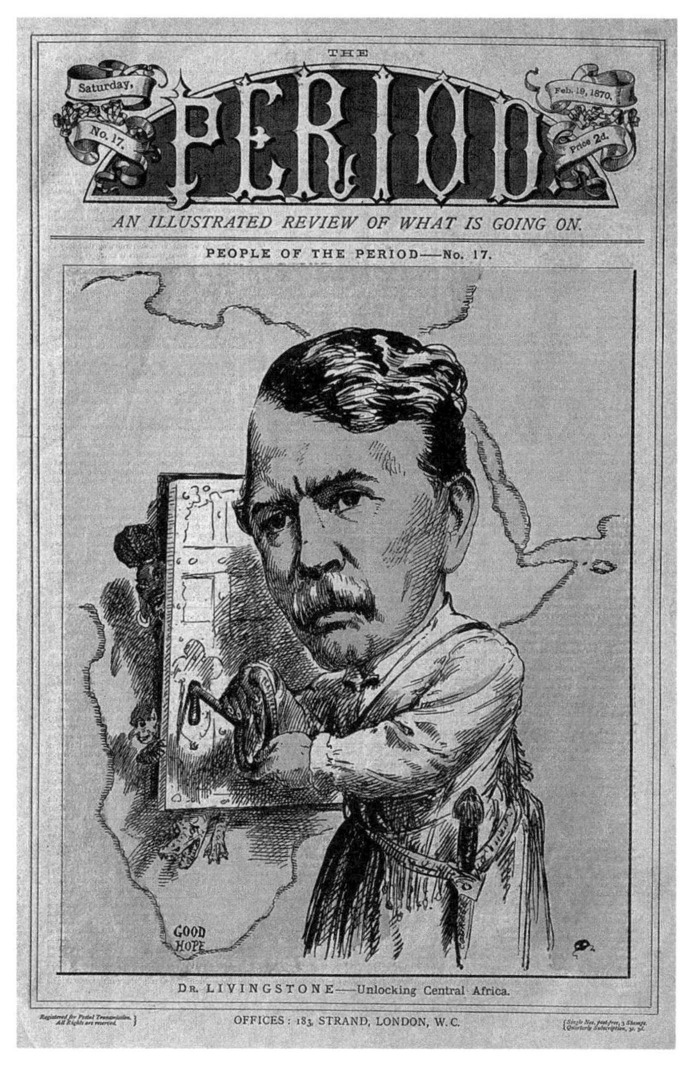

Portrait of Livingstone that appeared on the front of a popular publication in 1870, which is typical of the intense public interest in the explorer throughout the Victorian period, despite his almost permanent absence.

disenchanted government backers. His reception in Britain fell just short of disgrace as his once celebrated resilience was instead characterized as an unreasoned resolve that had indirectly caused the death of several on his expedition, as well as the demise of missionaries who had travelled to remote parts of Africa on his recommendation. Even so, just two years later he was again on his way to Africa, this time at the instigation of the Royal Geographical Society, with a new fixation: to settle the age-old mystery of the origins of the Nile.

Where Africa had not succeeded in breaking Livingstone's remarkable constitution, it did finally break his spirit. After five years of relentless hardship investigating possible sources of the Nile, he found himself destitute and reliant on the charity of the very Arab slave traders he had dedicated his life to supplanting. In 1871, four months before Stanley reached him, Livingstone had been in the remote market town of Nyangwe, in what is today the Democratic Republic of Congo. There he watched in horror the slaughter of more than 400 mostly women and children perpetrated by the same slave traders from whom he had accepted help and shelter for months. Appalled by the massacre – and at how he had been forced to compromise his principles – Livingstone fled. With feet ulcerated and swollen in tattered shoes and barely functioning under the torment of both extreme heat and chronic dysentery, he made his way through unfamiliar terrain to Ujiji. "The mind acted on the body and it is no overstatement to say that every step of between 400 and 500 miles was in pain," he recorded.

Stanley remained with Livingstone for five months, but when the time came to leave, Livingstone would not return with him to Britain, citing a need to finish his work and find the headwaters of the Nile. It is clear that this purely geographical question had become inseparable in his mind from the abolition of the African slave trade. Perhaps this fatal decision to continue also stemmed from a notion of redemption that ran beyond his sense of duty in his work as a mission from God. It is possible to imagine that finding the source of the Nile represented for Livingstone a way to atone for his acceptance of help from those involved in slavery and his powerlessness to stop the massacre at Nyangwe, to salvage his reputation as an explorer, and to justify the sacrifice of time with his family (something he described towards the end of his life as his only regret) as well as to expunge the grief and guilt over the death of his wife.

A little over a year after Stanley's departure, Livingstone died in today's Zambia, having continued on his quest even when bleeding profusely and needing to be carried in a litter. He still had one last extraordinary journey to make, however. Having buried his heart in African soil, three loyal companions carried Livingstone's embalmed corpse for more than 10 months and for 1,600 km (1,000 miles) across almost impassable terrain all the way to the coast. He was buried with full honours in Westminster Abbey and among his pall-bearers was Henry Morton Stanley – who at last did what he had promised and brought home the bones of the great explorer.

Livingstone's legacy is complex and controversial. A month after his death, his account of the bloody Nyangwe massacre precipitated a treaty between the governments of Zanzibar and Britain which effectively ended the Arabic slave trade network and realized Livingstone's lifelong ambition. Yet, in a cruel irony, his expeditions and those made by his protégé, Stanley, in the years that followed would facilitate the worst of the European colonialization of Africa and subject so many to an entirely new form of enslavement. We can only imagine that the spirit of Livingstone, as restless in death as he was resilient in life, still pervades the African wilderness in relentless pursuit of a better and more righteous future.

BEWARE THE RESILIENCE OF LIVINGSTONE

LIVINGSTONE FAMOUSLY SAID, "I AM PREPARED TO GO ANYWHERE, AS LONG AS IT IS FORWARD." HIS TENACIOUS DETERMINATION TO PERSEVERE IN THE FACE OF NEAR CONSTANT ILLNESS, FAILURE, SEPARATION FROM HIS FAMILY AND MYRIAD MENTAL AS WELL AS PHYSICAL BURDENS IS BOTH INSPIRING AND LAUDABLE, BUT HIS LIFE ALSO SERVES AS A WARNING OF THE PERILS OF A RESILIENCE THAT LACKS PERSPECTIVE AND DEMONSTRATES HOW EASILY RESILIENCE CAN TRANSFORM INTO UNHEALTHY OBSESSION.

Imagining of the meeting between Livingstone and Stanley taken from *The Life and Explorations of David Livingstone* (1875). This is representative of the contemporary depictions of the encounter that turned it into one of the most famous moments in the history of exploration.

RAISE UP YOUR THOUGHTS

Clear in Livingstone's mind was that his primary motivation for every action he took in Africa was the ambition to rid the continent of the trade in human lives. No matter the specific aims of any particular expedition, Livingstone linked these objectives back to his principal purpose – and this sustained his perseverance. Connecting a short-term aspiration to an enduring ideal is more likely to prompt greater resilience.

EMPATHY

Livingstone inspired notable loyalty from the African members of his expeditions, who responded to the care he took to understand their cultures, language and perspectives. This is in striking contrast to his expeditions involving Europeans, which were often marred by internal disputes and dysfunction. Livingstone lacked tolerance for anyone who couldn't match his own strength and resilience, making few allowances for any perceived weakness in his team, even when they fell prey to the inevitable tropical diseases. It resulted in frequently fatal mismanagement, and Livingstone is remembered as an erratic and unsympathetic leader.

INTEGRITY

Livingstone was willing to go to extreme lengths to achieve his purpose, but after witnessing the Nyangwe massacre, he perceived that he had fatefully compromised his principles in pursuit of an ideal. Resilience must be achieved with responsibility and integrity: the end will not justify the means.

INNOVATION

SYLVIA EARLE

STARING OVER AN ABYSS DURING ONE OF HER EARLY UNDERWATER EXPLORATIONS, SYLVIA EARLE WAS FRUSTRATED AT NOT BEING ABLE TO GO ANY FURTHER. "I WANTED TO GO AS DEEP AS THE OCEAN," SHE REMEMBERS. THE AMBITION HAS LED TO A LONG AND ACCOMPLISHED CAREER IN OCEAN EXPLORATION, ONE DISTINGUISHED BY HER INNOVATIVE USE OF NEW TECHNOLOGY TO ENABLE A GREATER APPRECIATION OF THE MARINE WORLD.

When Earle made her first enthusiastic dives beneath shallow seas at the age of 16, the only equipment available to her was a diving helmet. The bubble of air inside these cumbersome enclosures was supplied by a hose that kept the diver in constant contact with the surface. It was only in the early 1950s, when she started working toward a PhD in marine botany, that Earle was fortunate to have access to some of the earliest available self-contained underwater breathing apparatus (scuba). Earle was enthralled by scuba, describing it feeling like a "superpower". Completely untethered to the surface, she was free to explore for as long as she had air in the tanks on her back. This enabled her to observe the life that existed beneath the waves in situ for the first time, rather than as lifeless specimens in a lab. Earle quickly grasped how scuba could revolutionize our understanding of underwater ecosystems and was among the first to use it as a tool for conducting scientific research. When her PhD work exploring marine algae in the Eastern Gulf of Mexico was completed just over a decade later, it was considered groundbreaking. Today, it is still regarded as a landmark study, both for its original fieldwork approach using scuba and the fact that the extended duration of the work allowed

Earle shows a specimen to her colleague inside the Tektite habitat
during the two-week Tektite II Mission 6 in 1970.

significant changes in the region to be documented at a time when environmental change was not yet a prominent issue.

Despite her skill and passion for scuba, Earle would come to realize that even this seeming superpower had its limitations. In 1964, while still working on her PhD, Earle was invited to join the International Indian Ocean Expedition on the research ship *Anton Bruun* to explore the marine life in such exotic and remote locations as the Comoros, Aldabra, the Seychelles and the Farquhar Islands. The six-week journey required Earle to be parted from her two young children – aged two and four at the time – and she would be the only woman on the 70-strong team (a fact that, much to Earle's annoyance,

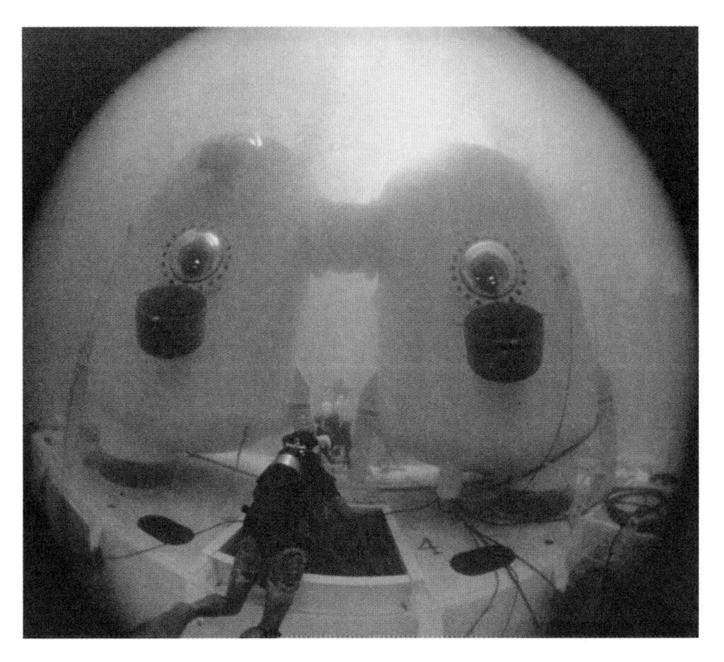

The Tektite habitat on the sea floor in the US Virgin Islands providing laboratory and living space for five team members.

caused more press interest than the many discoveries made by the expedition). Even so, this was her first opportunity to travel outside her native United States and she found the experience both inspiring and productive. Relishing every moment, she completed four further expeditions on the *Anton Bruun*, but was left with a nagging dissatisfaction. "We had a problem. We were in a boat trying to understand the great depths of the ocean from far above ... what would you know of New York City if you're flying overhead?" she asked. Earle was convinced that to truly understand marine life it was necessary to be up close and among it. In 1969, she seized an opportunity that would allow her to do just that.

Tektite was a futuristic, deep-sea laboratory secured on the sea floor at a depth of 15 m (49 ft) in Great Lameshur Bay off the island of St John in the tropical US Virgin Islands. Designed to maintain a constant internal pressure that was a little over twice the average air pressure at the surface, the habitat allowed divers to stay at depth for extended periods of time. Consisting of two tall cylindrical tanks connected by a narrow tunnel, the habitat featured laboratory and living spaces as well as a wet room where divers could access the surrounding ocean through a pool in the floor. The first Tektite mission involved four military-trained aquanauts living in the underwater base for 60 days. Earle, now a Harvard research fellow with more than 1,000 hours of dive experience, applied to join the second expedition, this time organized by NASA. Receiving submissions from eminently qualified female divers appears to have taken the organizers by surprise. "The people in charge just couldn't cope with the idea of men and women living together underwater," says Earle. Instead, she was asked to lead Tektite II Mission 6, a two-week expedition for which she would be living in the underwater habitat with four other female aquanauts. It was the first ever NASA mission of any kind to include women.

Mission 6, like all the Tektite expeditions, was primarily a study of human behaviour in a confined and hazardous environment, which would feed into wider investigations by NASA in preparation for future human space travel. For Earle, however, the appeal of Tektite was that it gave her the gift of time. Without the need to return to the surface, Earle and her fellow aquanauts spent up to 12 hours each day exploring the marine life that surrounded the Tektite habitat. They came to recognize not just endemic species of marine fauna and flora but individual fish and other marine animals resident in the immediate area. They became familiar with the routine, behaviour and even personality distinct to each individual animal. This new perception of the interconnectedness and sophistication of marine life was a defining revelation for Earle, but just a few years later, in 1979, came another encounter with technology so cutting-edge that it seemed like science fiction – and it would prove to be equally pivotal.

Bubbles "like blue champagne" filled Earle's view as she was lowered toward the sea floor. She was encased in a rigid, pressurized "suit" strapped to a small two-man submersible which descended for around half an hour to a site near Oahu, Hawaii, that had not previously been explored. The suit, looking like a stylized robot, had articulated legs allowing Earle some mobility, and arms fitted with claws that could be operated from within. When she touched down at a depth of 381 m (1,250 ft), the harness that held the suit in place against the submersible was released and Earle – with extreme effort – was able to "walk" across the seabed. The suit had been developed for industrial uses – such as marine salvage or the maintenance of underwater oil pipelines – but on hearing of the invention, Earle had engineered an expedition to trial the suit for scientific exploration. Her dive was the first time the suit had been used in the open ocean – previously it had always relied on connection to an air supply at the surface or had been lowered to the sea floor by a cable from a ship. Now, for two and a half hours, Earle explored the bottom of the ocean completely independent of the outside world except for a 5.5 m (18 ft) long communication cable, which connected her to the nearby submersible. Rescue from such a depth would have been near impossible, and without the protection of her suit, the pressure of the water would have been fatal. Regardless, Earle was wholly occupied, not by the danger but by the life she discovered in the deep. "There was a forest of corals that looked like giant bed springs. I concentrated on these bioluminescent spirals of black-and-white-banded bamboo coral (*Lepidisis olapa*) that bioluminesced when you touched them. They

Earle begins her descent to the sea floor in 1979 close to Oahu, Hawaii, in a pressurized diving suit that did not need a tether to the surface. It remains the deepest untethered dive ever made.

sent rings, like little glowing blue donuts of light shooting out from where you touched them." The dive remains the deepest ever made without a tether to the surface, but for Earle its greater significance was as the beginning of a lifelong fascination with the development of deep-sea technology for scientific exploration.

By the 1980s, Earle had become frustrated by what she saw as a lack of progress in deep-sea travel, especially when compared to other forms of exploration such as space, and bemoaned the lack of an "underwater equivalent of NASA" to push forward the needed technological development. Her solution was to found Deep Ocean Technology (followed by Deep Ocean Engineering and Deep Ocean Exploration and Research) to focus on the design, build and operation of submersibles and remotely operated vehicles (ROVs) specifically for exploration. On the board of the company were such luminaries of the underwater world as Jacques Piccard and Don Walsh – who, together, in the bathyscaphe *Trieste*, had been the first to dive to the deepest point of the world's oceans – as well as the astronaut Buzz Aldrin. Earle described her own role within the company to *The New Yorker* in 1989 as one of curiosity and stubbornness: "I would say, 'Why *can't* we just jump into the water and go down to the bottom of the ocean? We climb mountains, don't we? We go up in airplanes, don't we? We've even gone into space.' ... I was just stubborn enough to say, 'O.K., why can't we build vehicles to go seven miles down—the deepest known place in the sea?'" Urged by her insistence, the company produced Deep Rover, a one-person bubble-like submersible designed to go as deep as 1,000 m (3,300 ft) and capable of dives as long as a week, and Deep Flight, an underwater aeroplane for a single pilot lying prone, which could reach speeds of up to 15 knots at depths down to 1,220 m (4,000 ft). These innovations were not just about setting records or diving as deep as possible; the company strived to make it possible to come

Earle at the Californian offices of Deep Ocean Engineering in 1983 with prototype models of the deep-sea exploration submersibles the company was then developing.

closer than ever before to the undersea environment, even to interact with it. Robotic arms were created which, despite a heavy industrial appearance, could grasp and move an object as fragile as a chicken's egg without it being damaged.

Earle pioneered the new technologies on her expeditions – including a series of underwater explorations she led as part of the National Geographic Society's Sustainable Seas programme – creating records for solo submersible dives as deep as 1,000 m (3,280 ft) along the way. Innovation enabled her to make extraordinary discoveries, but also to witness first-hand the devastating change impacting the ocean over just a handful of decades. Conservation increasingly became her primary focus. Eschewing the common belief that then existed within academia that serious research should not mix with popular scientific communication, Earle sensed the importance of sharing what she had seen. In 2009, she founded Mission Blue with a very simple, yet powerful, new idea. Just as 12 per cent of land on the planet enjoyed some form of environmental protection, she wanted to create a global network of marine protected areas to safeguard the sea. She proposed doing this through Hope Spots – areas scientifically identified as critical to the health of the ocean, or what Earle refers to as "Earth's Blue Heart". At the time, less than 1 per cent of the world's seas had any form of protection. Now, after a decade in which a total of 131 Hope Spots have been created, Mission Blue has helped increase that number to 10 per cent and aims to increase it further to 30 per cent by 2030. Despite the challenges, Earle's innovative spirit refuses to be anything but optimistic about the future. "We have lost 50 per cent of the world's coral reefs. That means we still have 50 per cent of our reefs left and it's up to us to keep them alive and well," she points out. "There are plenty of reasons for hope."

After more than four decades at the forefront of ocean exploration, Earle has led more than 100 expeditions, logged over 7,000 hours underwater, and authored more than 190 scientific, technical, and popular publications. She has truly earned the sobriquet "Her Deepness", by which she is most often introduced today as she travels the world raising awareness and fighting the cause of conservation. Now an octogenarian, she shows no sign of slowing the frenetic pace of innovation that has marked her life. She celebrated her 80th birthday by snorkelling around the sea-ice edge of the Arctic Ocean, leading a group of artists and scientists as part of the Elysium Arctic project, and a recent expedition dive with a *National Geographic* journalist prompted him to write, "She seems to get younger as she gets deeper." It seems, even now, that Sylvia Earle remains as determined as ever to go "as deep as the ocean", whatever it might take.

Earle examines a patch of *Sargassum* seaweed in the Sargasso Sea close to Bermuda in 2010. This expedition was part of the Mission Blue campaign founded by Earle in 2009 and which continues work towards greater global ocean conservation.

DEMAND INNOVATION LIKE EARLE

AFTER A LIFETIME OF USING INNOVATION TO EXPLORE AND UNDERSTAND THE MARINE WORLD, SYLVIA EARLE WANTS US ALL TO TAKE ADVANTAGE OF OUR ABILITY TO INVENT A BETTER FUTURE THROUGH INGENUITY. SHE URGES, "IF YOU COULD CHOOSE A TIME TO BE AROUND, CHOOSE NOW. BECAUSE THIS IS THE FIRST TIME THAT WE KNOW THAT WE ARE CHANGING THE NATURE OF THE WORLD."

Earle during a 1970 research dive, collecting samples of plant life from the sea floor. She is trialling experimental rebreathing apparatus that enables a diver to stay underwater for longer.

THE IMPORTANCE OF OPTIMISM

Sylvia Earle has faced numerous and varied obstacles during her career, be they limitations preventing her own explorations or the challenges facing the preservation of the planet's oceans. Regardless of the scale or complexity of a problem, she approaches each barrier with the belief that we have the ingenuity as a species to innovate solutions and with a determined focus on hope.

BE A DISRUPTER

Unafraid to be challenging, Earle has never allowed herself to be limited by convention in either her personal or work life. Her dedication to following her passion is an example of how one person can demand and influence progress when they are willing to be the one that upsets the status quo.

NETWORK

Considering that she has spent most of her life at sea (or under it), it is astonishing that Earle manages to stay so well informed. Always abreast of the latest advancements in underwater technology, ocean exploration and marine research, she has a tireless ability to network and make connections. An avid collector of business cards and quick to pick up the phone, Earle has a rich web of contacts that she invests time and energy in maintaining – and as a result, her reach is broad and multifaceted.

PURPOSE

OLAUDAH EQUIANO

WHAT DIFFERENTIATES AN EXPLORER FROM A TRAVELLER IS THAT THE JOURNEYS OF AN EXPLORER HAVE PURPOSE. WHETHER THE OBJECTIVES ARE SCIENTIFIC, GEOGRAPHICAL, HUMANITARIAN OR SOMETHING ELSE, WHAT ALL EXPLORERS HAVE IN COMMON IS THAT THEY RETURN FROM THEIR TRAVELS WITH KNOWLEDGE THAT IS SHARED FOR THE GREATER GOOD.

In 1773, a young midshipman in the Royal Navy secretly ventured out onto the perilous pack ice of the Arctic Ocean in pursuit of a polar bear whose pelt he wanted to take home for his father. When the bear attacked and the midshipman's musket misfired, he was forced to use the firearm as a club to beat off the enraged bear. Spotted just in time from his ship, HMS *Carcass*, he was saved by cannon fire that frightened away the bear. The story is a famous one because the midshipman was a young Horatio Nelson, who would grow into the legendary British Admiral of Trafalgar fame. Less well known is that onboard the ship that accompanied the *Carcass*, HMS *Racehorse*, was an explorer who would have a very different but equally momentous legacy.

Both ships were part of a major Royal Navy polar expedition attempting to find the Northeast Passage, a sea route from Atlantic to Pacific via the Arctic. The expedition would ultimately be viewed by history as a failure. Not only did the two ships fail to find the Northeast Passage, but they also were unable to venture far beyond the already explored coasts of Spitsbergen, an archipelago north of Norway, before being trapped by sea ice. With little new discovered, they were forced to return to London barely four months after having set out. However, there was one achievement that caused a great deal of excitement.

Onboard *Racehorse*, the efficiency of new apparatus that could distil drinking water from seawater had been proven. This was the first time the idea had been trialled at sea. For a seafaring nation such as Britain, so heavily dependent on its naval and merchant fleet, the ability of ships to access drinking water was of critical importance, and the distilling apparatus was seen as a major technological breakthrough. The inventor was Dr Charles Irving, but its operation and testing on *Racehorse* during the expedition had been almost entirely undertaken by Irving's assistant, a man known as Gustavus Vassa.

Vassa was barely 30 years old when he departed for the Arctic and yet he was already a highly experienced sailor and as well travelled as anyone else aboard either the *Racehorse* or the *Carcass*. From an early age, he had sailed on a succession of Royal Navy warships, learning the skills of a life at sea and a life at war. When Britain and France fought the Seven Years' War, a global conflict that gradually embroiled all the major European superpowers of the day, Vassa experienced some of the key military action in both North America and the Mediterranean as an able seaman. He witnessed the confusion and brutality of naval confrontations from the gundecks of frigates and flagships, charged with hauling gunpowder amid the smoke and carnage of battle in rough seas, while the ship made wild manoeuvres to fatally close in on prey or make desperate attempts at escape. After the war, Vassa spent many more years sailing trade routes in the Americas and across the Mediterranean as far as Turkey, able to indulge his curiosity for exotic cultures and new experiences. His travels provide a fascinating insight into the reality of eighteenth-century life, but what makes Vassa stand out among his contemporaries – as it still does today in the histories of exploration – was his perspective as a non-European. Vassa had been born in a remote part of modern-day Nigeria, the son of a tribal chief. Traumatically kidnapped into slavery as a boy and shipped from Africa to North America, he spent the first decades of his life being considered the property of others, passed from owner to owner with no rights whatsoever over his fate or his well-being. Even the name he used, Gustavus Vassa, was a whimsy imposed on him by one of the first men to buy him. His real name was Olaudah Equiano and, uniquely, he would become an explorer not by choice but by circumstance.

It was as the slave of a British Royal Naval officer that Equiano first began to ascribe a purpose to his travels. He had been not more than 10 years old when the officer purchased him from a tobacco plantation in Virginia, then a British colony. For eight years, Equiano worked as the officer's servant on a succession of Royal Navy ships travelling to all corners of the Atlantic. Initially terrified of the sea and of the Europeans he had good reason to fear, Equiano decided to approach his unwanted voyaging as an "opportunity to better myself". It became his own project, exploring and discovering the European culture around him – much as later Victorian anthropologists from Europe would come to study Equiano's native Africa. He learned the seemingly mysterious art of "hearing books speak" (reading) and became a skilled seaman, mastering elements of celestial navigation. He drew comfort from the feeling of control his sense of purpose gave him over his vulnerable circumstances, and noticed the effect it had of lessening his fear. In fact, he realized he was moving to the opposite extreme, finding that nothing scared him anymore, to the extent that he yearned for military action at sea, wanting more fear, more experience. "I longed to engage in new adventures and see fresh wonders," he wrote.

Equiano had a talent for observation on his enforced travels, particularly in his attention to detail. He commented, "I had a mind on which everything uncommon made its full impression." It's a characteristic shared by many of the greatest explorers, the ability to absorb experiences and recognize the significance in even the seemingly mundane, noticing evocative details that can inspire greater understanding. But experience by itself is not exploration. The information gathered must be shared in order for it to provide purpose and meaning. The ability to communicate what you have seen in a way that allows others to learn is vital. Fortunately, Equiano was a natural storyteller. He relates that even in the chaos of a fierce naval battle against the French, on board HMS *Namur* in the straits of Gibraltar towards the end of the Seven Years' War, he was "pleasing myself with the hope, if I survived the battle, of relating it and the dangers I had escaped". Despite not yet knowing when or how he might have the opportunity to share his experiences, Equiano intuitively stored up everything he saw. This instinctive recognition of the importance of sharing what he had witnessed and his skill as a communicator would drive forward his sense of purpose throughout his life and distinguish him as an explorer.

When Equiano was abruptly sold and shipped to the West Indies, his abilities as a sailor were put to use on cargo vessels, which allowed him to start making some small, but astute, trades of his own, even though this put him at great risk of physical punishment. After four years of careful effort, plagued by bitter disappointments and catastrophic setbacks, he was eventually able to triumphantly buy his freedom. At the age of 21, Equiano was no longer a slave, but he quickly discovered his position in society as a free man of colour was dangerously precarious. Subject to violent racism and overwhelming injustice often enshrined in law, Equiano would charge his travels with an ambitious new purpose: the abolition of slavery.

In 1789, having settled in London, he published his autobiography, *The Interesting Narrative of the Life of Olaudah Equiano, or Gustavus Vassa, The African.* The book detailed what he remembered of his Igbo heritage in Africa, including descriptions of the scarification practices that uniquely differentiated his culture, and treasured memories of the family and life that had been stolen from him. But the importance of Equiano's skill as a storyteller really comes to light in his descriptions of enslavement and his experiences traveling as a free man across Europe and the Americas. It is impossible to

The seal of the Society for the Abolition of the Slave Trade, founded in England in 1787. The Society promoted Equiano's autobiography and invited him to lecture widely around the country.

overstate the galvanizing effect this first-hand testimony, delivered with such intimate detail, produced on a nascent abolitionist sentiment among the British public. The movement against slavery and especially the transatlantic slave trade had only just begun to emerge in Britain, led by small pockets of academics, politicians and Quakers struggling to raise awareness of the cause in a country which, although benefiting economically from slavery, was largely oblivious to its reality. Equiano's vivid accounts of the transport

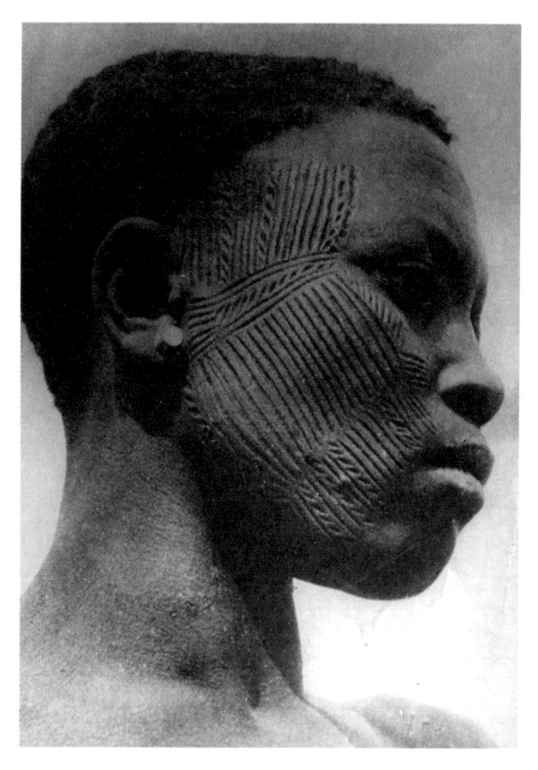

Portrait (c.1921) showing *ichi*, the ritual facial scarification worn by the Igbo people to denote the passing of initiation rites and status as nobility.

of kidnapped Africans across the Atlantic established what became known as the Middle Passage as an atrocity in the public consciousness, revealing horrors that had previously been unknown. His expert use of language is what made his account so affecting and, today, some 230 years after it was written, it remains a harrowing read. It is not just the acts of cruel violence that Equiano records or the dehumanization and degradation of an entire race of people that is so devastating – facing the reality that humankind is capable of inflicting such horror on itself is equally sickening. Equiano held up a mirror so that eighteenth-century Europe saw itself through African eyes for the first time – and was horrified by what it saw.

The book alone wouldn't have had such influence were it not for Equiano's determination. He travelled extensively across England, Scotland and Wales in the years that followed publication, proving a tireless and effective campaigner, evidently as good a communicator in person as he was on the page. His lectures inspired impassioned public support for the abolitionist cause, while private meetings with prominent politicians proved equally persuasive. He became a significant public figure in Britain: the arguments and evidence collected against slavery throughout his travels were read in Parliament, presented to royalty and regularly published in national newspapers, while his book reached nine editions in his lifetime, including an American edition published in New York.

Equiano did not live to see the abolition of slavery, but when the Slave Trade Act of 1807 was passed some 10 years after his death, followed by the Slavery Abolition Act of 1833 that brought a decisive end to slavery throughout the British Empire, his crucial role in the abolition movement was rightfully acknowledged.

Equiano's ambition to end slavery must have appeared fanciful and unlikely when he had first conceived his purpose. At that time, not only had he been a man with the lowest status in society, deprived of basic rights and utterly powerless, but slavery, the object of his ambition, was also a vast industry of invincible wealth protected by powerful governments and commerce. Yet the child from Igbo and former slave not only achieved his ambition, he also provided a legacy and an example that remains inspiring and relevant centuries later. The life of Olaudah Equiano is remarkable in many ways, most particularly because it shows us that no matter what our circumstances might be, however powerless we feel or however monumental the task at hand, with patience, dedication and purpose we have the chance to make a difference every day of our lives.

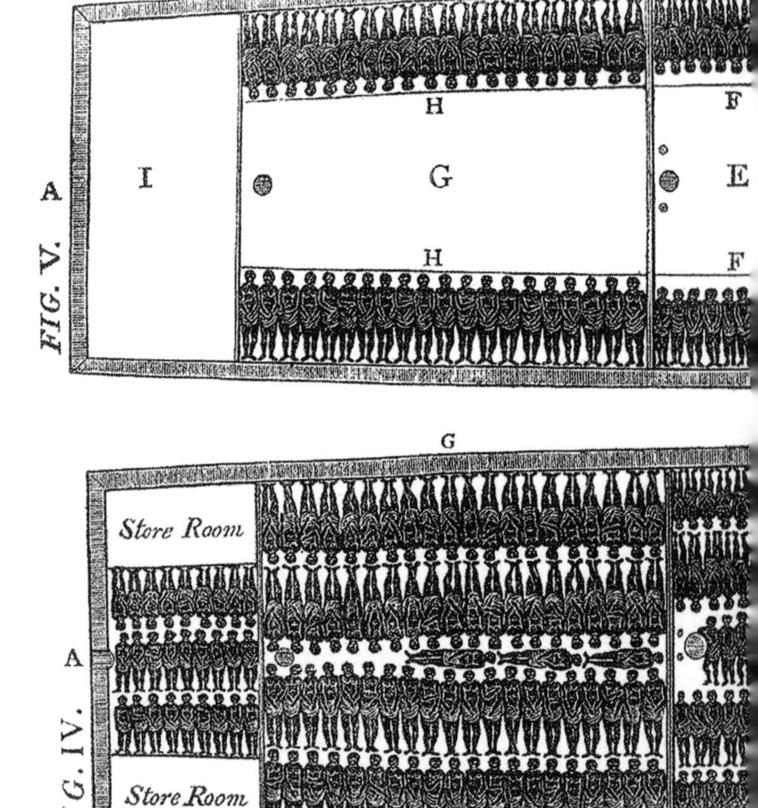

Diagram of an Atlantic slave ship showing the typical loading of slaves as cargo across what became known as the Middle Passage. The diagram was included in material presented to Parliament by abolitionists in 1791.

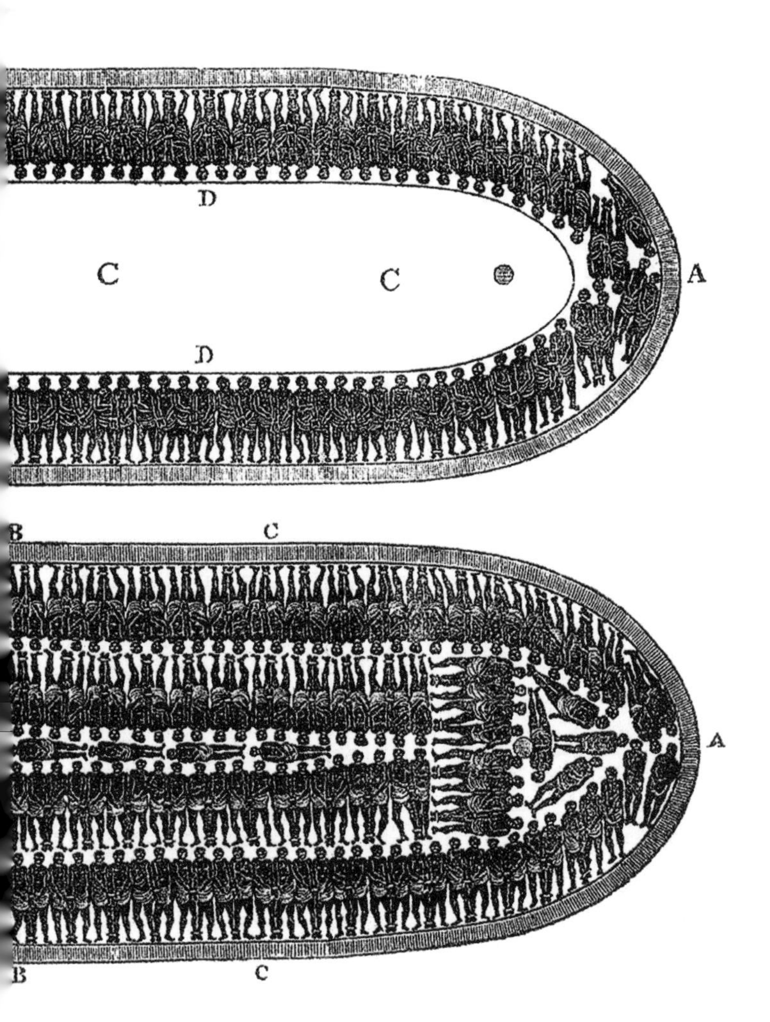
D
C
C
A
D
B
C
A
B
C

FIND PURPOSE LIKE EQUIANO

GREATER PERSONAL FULFILMENT AND REWARD IS ACHIEVED WHEN WE HAVE PURPOSE. HOWEVER, FINDING A SENSE OF PURPOSE AND FOLLOWING IT THROUGH CAN OFTEN BE A CHALLENGE. ASCRIBING PURPOSE TO JOURNEYS IS A FUNDAMENTAL TENET OF EXPLORATION, SO WE CAN LOOK TO EXPLORERS LIKE OLAUDAH EQUIANO FOR ENCOURAGEMENT.

An engraving of the attack and capture of Belle Île by the British in 1761 during the Seven Years War, a battle in which Equiano participated aboard HMS *Namur*.

START WITH YOURSELF

As a slave, Equiano had nothing. Yet, even with nothing, he still had his story, his experiences and the innate talents unique to him. In searching for purpose, the best place to start is with ourselves. Drawing on our own experiences to establish what it is that we care about deeply, can be used as a foundation on which to develop clear purpose. Identify your innate skills as an individual and build on them as your primary tools for succeeding in whatever your chosen purpose might be.

COMMUNICATION

Equiano's mastery of written and verbal communication and his ability as a storyteller were key to his achievements. In driving forward purpose, it is important to define a clear narrative and share experiences in ways that amplify your voice and engage others, be it through art, science, literature or a medium unique to you and your skills.

BEGIN

It is very easy to find justifiable reasons to procrastinate, easy to convince ourselves that it is prudent to wait for a particular set of circumstances to ensure success. The truth is, there is no such thing as a perfect moment to begin. The story of Equiano demonstrates that, no matter how inopportune our situation, there is always valuable progress to be made. Make a start.

NOTES

1. Christie Mallowan, A. (1946) *Come Tell Me How You Live: An Archaeological Memoir*. Glasgow: William Collins, Sons

2. Christie, A. (1937) *Death on the Nile*. Glasgow: William Collins, Sons

INDEX

Caption pages are in italics

H

I

J

K

L

M

N

O

CREDITS

The publishers would like to thank the following sources for their kind permission to reproduce the pictures in this book.

Akg-images: Fratelli Alinari 68; /Pictures from History 196-197; /TT News Agency 120

Alamy: AF Archive 85, 86; /Alpha Stock 188; / Chronicle 75; /Classic Photographics 72; /Echo Images 132; /Everett Collection Historical 70; / Alain Le Garsmeur 183; /Iconographic Archive 126; /Newscom 119; /Retro AdArchives 98; /Science History Images 151

Bridgeman Images: Everett Collection 193; / Natural History Museum, London 141; /Royal Geographical Society 142

Getty Images: 44; /Thomas Baines/Royal Geographical Society 166; /Gertrude Bell/Royal Geographical Society 89, 92; /Bentley Archive/ Popperfoto 96; /Bettmann 8, 38, 99, 102-103, 108, 174, 177, 180-181, 186; /Luis Dafos 65; /Nicolas Le Corre/Gamma-Rapho 79; /DeAgostini 157; / Heritage Arts/Heritage Images 106; /E.H. Hewitt/ Library of Congress/Corbis/VCG 10; /History and Art Collection 137; /Frank Hurley/Scott Polar Research Institute, University of Cambridge 51, 54, 56; /G.I. Davys/Royal Geographical Society 61; /Hulton-Deutsch Collection/CORBIS 49; / Keystone 18; /Keystone-France/Gamma-Keystone 35; /Keystone/Hulton Archive 110; /L.A. Wallace/ Royal Geographical Society 162-163; /Mansell/ The LIFE Picture Collection 94; /Mondadori 15; / Photo12/UIG 13; /The Print Collector 46, 128-129, 134, 138; /Royal Geographical Society 58, 90-91; / Shaul Schwarz 185; /Tetra Images (endpapers); / Topical Press Agency 36; /Touring Club Italiano/ Marka/Universal Images Group 64; /Underwood & Underwood/CORBIS 52; /Universal History Archive/ Universal Images Group 158; /J. Weston & Son/Royal Geographical Society 83; /ZU_09 123; /Martin Zwick/REDA&CO/Universal Images 125 (bottom)

© Hubertl Wikimedia Commons CC BY-SA 4.0: 148

Jaan Künnap: 114

Library of Congress, Washington D.C.: 37

NASA: 20, 23, 24, 28-29, 30

Natural History Museum Images: Trustees of the Natural History Museum, London 146
National Oceanic and Atmospheric Administration: 176

Courtesy Princeton: 66-67

Public Domain: 112, 145, 154-155, 198

Preus Museum co Paul Berge Roald: 16

Private Collection: 194

Shutterstock: AP 42, 43, 80; /Keystone Pictures USA: 32; /Oli Scharff 101; /Tabei Kikaku Co Ltd/AP 116; /Umomos 125 top

Wellcome Collection. Attribution 4.0 International (CC BY 4.0): 160, 169, 172

Every effort has been made to acknowledge correctly and contact the source and/or copyright holder of each picture and Welbeck Publishing Group apologises for any unintentional errors or omissions, which will be corrected in future editions of this book.

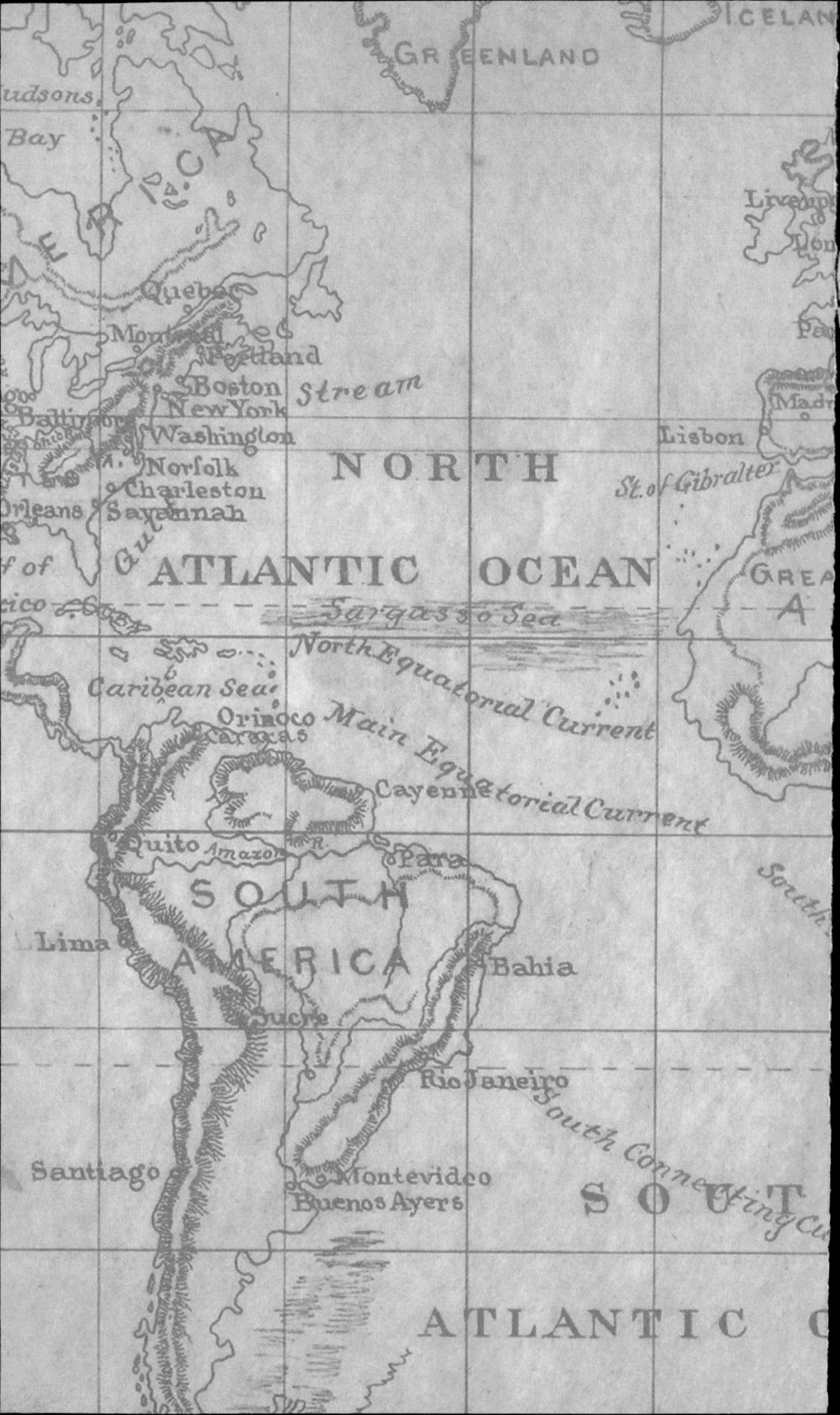

GREENLAND

ICELAN

Hudsons

Bay

AMERICA

Liverp

Lon

Par

Madr

Quebec

Montreal

Portland

Boston

New York

Washington

Norfolk

Charleston

Savannah

Orleans

Stream

Gulf

NORTH

ATLANTIC OCEAN

Lisbon

St. of Gibralter

GREA

A

Sargasso Sea

North Equatorial Current

Caribean Sea

Orinoco

Caracas

Main Equatorial Current

Cayenne

Quito

Amazon R.

Para

SOUTH

AMERICA

Lima

Bahia

Sucre

Rio Janeiro

South Connecting Cu

Santiago

Montevideo

Buenos Ayers

SOUT

ATLANTIC